CAMBRIDGE LIBRARY COLLECTION

Books of enduring scholarly value

Earth Sciences

In the nineteenth century, geology emerged as a distinct academic discipline. It pointed the way towards the theory of evolution, as scientists including Gideon Mantell, Adam Sedgwick, Charles Lyell and Roderick Murchison began to use the evidence of minerals, rock formations and fossils to demonstrate that the earth was older by millions of years than the conventional, Bible-based wisdom had supposed. They argued convincingly that the climate, flora and fauna of the distant past could be deduced from geological evidence. Volcanic activity, the formation of mountains, and the action of glaciers and rivers, tides and ocean currents also became better understood. This series includes landmark publications by pioneers of the modern earth sciences, who advanced the scientific understanding of our planet and the processes by which it is constantly re-shaped.

Life, Letters, and Works of Louis Agassiz

On the basis of extensive material in the form of letters, pamphlets and the recollections of friends and contemporaries, Jules Marcou (1824–1898) tells the story of the life and work of Louis Agassiz in this two-volume work of 1896. The Swiss-born palaeontologist, glaciologist and zoologist (1807–1873) is regarded as one of the founding fathers of the modern American scientific tradition. Marcou, a fellow countryman and collaborator of Agassiz, does not attempt to conceal his high regard for the subject of his biography but does have 'in view the truth'. In a chronological narrative, Volume 1 traces the childhood and early professional success of Agassiz, including his charming of the great von Humboldt. It describes Agassiz' time as professor in Switzerland and his marriage, ending with the arrival of Agassiz in America and his first attempts at forging a university career there.

Cambridge University Press has long been a pioneer in the reissuing of out-of-print titles from its own backlist, producing digital reprints of books that are still sought after by scholars and students but could not be reprinted economically using traditional technology. The Cambridge Library Collection extends this activity to a wider range of books which are still of importance to researchers and professionals, either for the source material they contain, or as landmarks in the history of their academic discipline.

Drawing from the world-renowned collections in the Cambridge University Library, and guided by the advice of experts in each subject area, Cambridge University Press is using state-of-the-art scanning machines in its own Printing House to capture the content of each book selected for inclusion. The files are processed to give a consistently clear, crisp image, and the books finished to the high quality standard for which the Press is recognised around the world. The latest print-on-demand technology ensures that the books will remain available indefinitely, and that orders for single or multiple copies can quickly be supplied.

The Cambridge Library Collection will bring back to life books of enduring scholarly value (including out-of-copyright works originally issued by other publishers) across a wide range of disciplines in the humanities and social sciences and in science and technology.

Life, Letters, and Works of Louis Agassiz

VOLUME 1

JULES MARCOU

CAMBRIDGE UNIVERSITY PRESS

Cambridge, New York, Melbourne, Madrid, Cape Town, Singapore,
São Paolo, Delhi, Dubai, Tokyo, Mexico City

Published in the United States of America by Cambridge University Press, New York

www.cambridge.org
Information on this title: www.cambridge.org/9781108072601

© in this compilation Cambridge University Press 2011

This edition first published 1896
This digitally printed version 2011

ISBN 978-1-108-07260-1 Paperback

This book reproduces the text of the original edition. The content and language reflect the beliefs, practices and terminology of their time, and have not been updated.

Cambridge University Press wishes to make clear that the book, unless originally published by Cambridge, is not being republished by, in association or collaboration with, or with the endorsement or approval of, the original publisher or its successors in title.

LIFE, LETTERS, AND WORKS
OF LOUIS AGASSIZ

Louis Agassiz
1872

LIFE, LETTERS, AND WORKS

OF

LOUIS AGASSIZ

BY

JULES MARCOU

WITH ILLUSTRATIONS

VOL. I

New York
MACMILLAN AND CO.
AND LONDON
1896

All rights reserved

Copyright, 1895,
By MACMILLAN AND CO.

Norwood Press
J. S. Cushing & Co. — Berwick & Smith
Norwood Mass. U.S.A.

PREFACE.

MORE than twenty years have passed since the death of Louis Agassiz, and although many biographies were published directly after his death, no true life of him has yet appeared: nearly all have been too eulogistic, while, on the other hand, some rather severe strictures and criticisms have incidentally appeared in articles purporting to give the life of some of his associates, or dealing with some special questions of natural history. Agassiz occupied too large and important a place in natural history not to have left both a certain number of critics and a larger number of enthusiastic admirers ready to see in him only faults or perfection. Truth lies between the two. As one of the reviewers of Agassiz's life by his wife says: "The true history of Agassiz has not yet been written."

To meet this want, I have made during the last twenty years a large collection of material in the form of letters, recollections of friends and contemporaries, and rare pamphlets, with the design of presenting to the public the man himself; his origin, his character, his public life, his private life, his passions, his weaknesses, his faults, his errors, his genius; what he did and what he left undone; above all, to put him in his place, in a true light, in correct perspective, with its lights and shadows, in the field of the history of natural science.

I have tried to speak of him uninfluenced by the discordant voices which have celebrated his merits without discretion, or demolished his reputation without measure. His faults were small, while his genius was great. "Son envergure immense allant d'un bout à l'autre du ciel scientifique," as was said also of Humboldt and Cuvier.

I enjoyed his friendship during almost thirty years, being one of the few men to whom Agassiz half unbosomed himself; and I am the last survivor of the small band of European naturalists who came to America with him. My admiration of the man is not concealed; but I have had constantly in view the truth, and have tried to be just, not only towards him, but also towards all those who were more or less connected with him during his scientific life.

In the thought of many, a man of genius ought to be perfect; and consequently when errors, mistakes, and faults appear, it is difficult to accept them and bear them with equanimity and indulgence.. But we must be generous, and make a fair allowance for human weakness, even in a man of genius, and especially in a man of genius.

Agassiz kept up all his life a very large correspondence, either directly, or when too busy or in ill health, by dictation. In Neuchâtel he wrote at least five letters daily, not only to naturalists, savants in general, and to his relatives, but also to other friends, and even statesmen and historians like Thiers and Guizot in France, and later to Dom Pedro II, in Brazil. The number of his letters is enormous, and until 1842 he kept copies of them all. I know of one of his correspondents who received more than one hundred letters from him. To choose among them is not an easy task. Mrs. Agassiz, in the life of her husband, has given a certain number (about ninety), selecting more especially those addressed to Agassiz's mother, father, and brother, and to some well-known men of science, philosophers, philanthropists, and politicians; besides giving letters written to Agassiz by naturalists like Humboldt, Cuvier, Buckland, Sedgwick, Lyell, etc. Unwilling to repeat what has been already so well done by Mrs. Agassiz, my quotations are limited to letters of Agassiz, addressed to practical naturalists, his contemporaries, working on kindred subjects. To see and appreciate the influence exerted by Agassiz on the progress of palæontology, geology, and the glacial question, it is important to show his impressions at the very moment when he received them in the course of his studies.

I have received much information, and copies of letters and notes, from persons or families formerly in correspondence with Agassiz. I beg them to receive my thanks; and I have especially to thank my good friend, M. Auguste Mayor, of Neuchâtel, first cousin of Agassiz. Although some years younger than Louis Agassiz, he knew him as a very young man, and followed closely his eventful and splendid career during his whole life, both in Europe and in America, for M. Mayor lived for more than twenty years in Brooklyn, New York, and received Agassiz at his house when he came to the New World. Each had perfect confidence in the other; as cousins and friends they loved one another without reserve. For myself, I cannot separate Louis Agassiz from Auguste Mayor. Such friendly and constant relations in both hemispheres between two men are extremely rare.

I have carefully read and considered all documents, and have made constant use of my intimate knowledge of Agassiz. Scientifically we did not agree on all points; but both were satisfied to accept our differences of opinions. On the whole, our friendship was never shadowed by a single serious disagreement.

The biography of such a man as Agassiz cannot be given by the publication of his letters only; because in letters the confidences are not so free, precise, or so full as can be desired: besides, many letters, for various reasons, cannot be published in full. Agassiz's genius was so spontaneous, so frankly natural, so absolutely sincere, that his physiognomy was most attractive, showing always the great mobility of his sentiments. He was one of those very few men whose works are not sufficient to make him entirely known; one must meet him face to face. Agassiz was so full of personal inspiration and original thought, that in order to have a just idea of him, naturalists went to Neuchâtel, and afterward to Cambridge, only to see him, and shake his hand. His individuality was a subject of continual observation by all those who surrounded or approached him. He was of an extremely rare and very complex type. It is impossible to group round him other naturalists, and to form a special class of spirits related to his. He surprised every one by his constant watchfulness, and his quickness to get at the truth of nature. Agassiz himself was more interesting than his works. His life is a rare study.

Until 1838, he wrote with his own hand an enormous amount of manuscript; nothing discouraged him, and he was always ready to use his pen, even to copy papers or books which he was too poor to purchase, or which it was impossible to procure otherwise. He kept a private journal, in which he wrote with great *naïveté* everything which occurred to him, or came under his eyes, when at Bienne, Lausanne, Zürich, Heidelberg, Orbe, Munich, Concise, Paris, and during the first two years of his life at Neuchâtel. He showed me this journal, and I had the privilege of reading in it some of his student adventures and escapades. I do not know what has become of the manuscript.

Agassiz, from his youth until his last illness, was overflowing with intellectual spirit and vitality. He is a rare example of manly qualities and activities. His influence on the progress and diffusion of natural history is second to none.

I have tried to bring him before the reader as I have known him. If I do not produce an exact portrait of the man and his life, it is due simply to my inability to express my feeling for the man and his works. Born at the foot of the Jura Mountains like Agassiz, and not far from his birthplace, I passed my youth and was educated under much the same circumstances as he, and ought to be able to deal with the difficult task of writing his biography, for I have had unusual opportunities to know him and his surroundings in the Old and New Worlds. I can truly say that the task of writing his life has been a work of friendly love and respect for the man, and of justice to the savant.

My aim has been constantly to make a judicious blending of history, correspondence, and extracts from his works, and of the estimation in which these are held by others. Mrs. Agassiz's account of her husband's life gives the character of Agassiz by means of a list of qualities rather than a complete picture. As is very likely to happen, her biography is rather a panegyric than an analysis of character. I have her example constantly before my eyes, in my endeavour not to fall into the same error; as Massimo d'Azelio says: "I must be honest, not only with the reader, but with myself; otherwise I should be treating the life of Agassiz like a half-decayed peach, the spoilt part of which I should cut out, and present only the sound portion."

Without passing over in silence the moral failings of the man, and the inequalities of talent of the naturalist, I have expressed all my admiration for this master of natural history. The unity which is not to be found in his acts or in his works will be found in his iron will; he had a fixed idea — he wished to be the "first naturalist of his time," as he said in a letter to his father, when he was still a student at Munich.

Although Agassiz was very ready to care for his own interests, he never was a practical man, in the full business sense of the word. Unable to choose suitable men as assistants and co-workers, he was very prompt to make use of them, whenever they were competent.

I shall finish with a French sentence, very appropriate to the Franco-Swiss savant, educated as a naturalist in Germany, a constant admirer and pupil of Cuvier, and finally a naturalized American. Agassiz "restera une personnalité populaire et sympathique. À mesure que ses défauts et ses faiblesses diminuent dans l'éloignement, ses qualités maitresses apparaissent plus éclatantes et font oublier tout le reste: il avait la foi, la vie, la chaleur, l'enthousiasme, la passion, et surtout ce qui le rendait éminement sympathique, il ne connaissait pas le fiel, l'envie, la rancune et la haine."

CAMBRIDGE, MASSACHUSETTS,
March, 1895.

INTRODUCTION.

FRENCH was the native tongue of Louis Agassiz. His remarkable and admirable mother knew neither English nor German, but wrote French with great purity and choice of expression. As one of the family in Switzerland writes me, "ses lettres sont charmantes, elle écrivait à merveille." All the great works of Agassiz, on which his reputation as an original naturalist is based, are in the French language. The most active part of his life, as regards great discoveries, was spent at Neuchâtel, then a small town, where French is the only language spoken. Before he came to America, all his correspondence with English naturalists was in French; so that it is almost impossible entirely to suppress this language in writing an accurate and true life of him. Translations, however good, never give an exact idea of what the author means, especially in the case of difficult and delicate observations in natural history.

After long consideration, I have, therefore, concluded to give what I quote of his correspondence in the original. All English-speaking naturalists read French now. As the interest of such a book is limited to naturalists and persons whose education leads them to read and appreciate the life of an extraordinary man, a few letters and quotations in French, scattered through the work, will present no difficulties to its readers. On the contrary, they will be a sort of stimulant and a relish, both scientifically and from a literary point of view.

For the same reason, an address — the most important of the many delivered by Agassiz — has been reproduced in the original. It is his celebrated discourse of 1837, on a "Great Ice-age."

During the period from his twenty-third to his fortieth year, Agassiz wrote many letters in German, mainly of a private nature, addressed to members of his German family or to a few most intimate friends. Almost all his scientific letters were directed to his old professor at Heidelberg, H. G. Bronn, and have been published in the "Neues Jahrbuch für Mineralogie, Geologie und Petrefaktenkunde." Although he was a perfect master of German, speaking and writing it like a native of Heidelberg or Munich, he never published important papers in that language, only a few pamphlets, the principal one being his reply to Karl Schimper's claims, 4 pp. 4to.

Agassiz's remarkable personality cannot be properly understood without taking into account the strength of his French nature. A Franco-Swiss he was born; and a Franco-Swiss he remained all his life, notwithstanding his American naturalization and his great admiration of the New World in general, and of the United States in particular.

CONTENTS

CHAPTER I.

1807–1827.

CHAPTER II.

1827–1831.

CHAPTER III.

1831–1832.

CHAPTER IV.

1832–1835.

CHAPTER V.

1836–1837.

CHAPTER VI.

1836–1837 (*continued*) and 1838.

CHAPTER VII.

1839–1840.

CHAPTER VIII.

1841–1842.

CHAPTER IX.

1843–1844.

CHAPTER X.

1845.

ILLUSTRATIONS.

VOL. I.

CHAPTER I.

1807–1827.

ANCESTRY — ORIGIN OF THE NAME AGASSIZ — COAT-OF-ARMS — BOYHOOD — MOTIER-EN-VULY — REPUTATION OF LOUIS'S FATHER AS A TEACHER — COLLEGE OF BIENNE — VINTAGE-TIME AT MOTIER — SOME PECULIARITIES IN THE CHARACTER OF LOUIS AGASSIZ — COLLEGE STUDIES AT LAUSANNE — HIS RESOLUTION TO BE A NATURALIST — UNIVERSITY OF ZÜRICH — HIS FIRST TEACHER OF ZOÖLOGY, M. SCHINZ — "FIRST AT WORK AND FIRST AT PLAY!" — UNIVERSITY OF HEIDELBERG — ALEXANDER BRAUN, KARL SCHIMPER, AND AGASSIZ — FIRST VISIT TO THE BRAUN FAMILY AT CARLSRUHE — TYPHOID FEVER — HIS STAY AT ORBE.

THE Agassiz family came originally from Orbe and the small village of Bavois, in the "Jura Vaudois." A little west of Orbe there is a small hamlet, called Agiez or Agiz. In old French, and more especially in the patois of the Canton de Vaud, Agiz, Agiez, Agasse, Agassiz, and Aigasse[1] mean "magpie," a bird which was and is still very common in the country around Orbe and La Sarraz. In low Latin, magpie is Agasia; in Provençal, Agazia, or Agassa, and Agasse; while in Burgundy and Franche-Comté it is Aiguaisse. Obvi-

[1] "Agassiz ou Agasses ou Agaisse; dans toute la France ces trois noms signifient Pie. Autrefois Agassiz s'écrivait un peu diffèrement Agacie. C'était un surnom donné jadis aux querelleurs, dit-on, et aussi aux grands causeurs. — On sait combien la Pie est jaseuse." (See "Dictionnaire des noms," par Lorédan-Larchey, p. 4, Paris, 1880.)

ously the name of the family was derived from the name of the bird; as evidence of this, the old armoiries or coat-of-arms of the Agassiz family is a black magpie on a silver ground (*Pie noire sur fond d'argent*), a drawing of which may be seen on the title-page. It is still preserved in the family in Switzerland, which also possesses an old seal, engraved on copper, with the same bird in the centre. Formerly, among all French-speaking peoples it was the custom for ennobled burghers to adopt for their coat-of-arms what was called "armes parlantes," and the Agassiz of Orbe chose the magpie.

Originally, the name was doubtless given to one inclined to talk a little too much, — as the French proverb has it, "bavard comme une pie."

One of the most faithful correspondents and best friends of Agassiz, Sir Philip de Grey Egerton, the great English paleoichthyologist, an excellent French scholar, used often to call him "Mon cher Agass" as a reminder of his knowledge of old French and patois.

The name Agassiz is not very rare, and is found among French people not connected in any way with the original family of Louis Agassiz. However, a branch of his family emigrated to London, and some fifty or more years ago a family of bankers of the name of Agassiz was there, who occasionally corresponded with their relatives of the Canton de Vaud. One of them published, in 1833, a book of travels under the title, "Journey to Switzerland and Pedestrian Tours in that Country" (London, 8vo), a work which is sometimes wrongly attributed to Louis Agassiz.

At the beginning of this century there was in Paris a M. Agasse, a publisher and bookseller, who published in 1804 the third edition of "La Flore française," by Lamarck and De Candolle.

The celebrated La Fontaine, in his fable "L'Aigle et la Pie," says: —

> "L'aigle, reine des airs, avec Margot la pie,
> Différentes d'humeur, de langage, et d'esprit,
> Et d'habit,
> Traversaient un bout de prairie.
> Le hazard les assemble en un coin détourné.
> L'Agasse[1] eut peur; mais l'Aigle ayant fort bien diné,
> La rassure, et lui dit: 'Allons de compagnie;
> Si le maître des dieux assez souvent s'ennuie,
> Lui qui gouverne l'univers,
> J'en puis bien faire autant, moi qu'on sait qui le sers.
> Entretenez-moi donc, et sans cérémonie.'
> Caquet-bon-bec alors de jaser au plus dru,
> Sur ceci, sur cela, sur tout. L'homme d'Horace,
> Disant le bien, le mal, à travers champs, n'eut su
> Ce qu'en fait de babil y savait notre Agasse."

The name is also found in Italy, but omits the *z* at the end.

In the Arabic or Mauresque and Saracenic language the expression "Kol-Agassiz" means *wing-leader*, a sort of field officer occupying a position between captain and major, called in Turkish "Bin-Bashi." So in Arabic Agassiz means *conductor*, *leader*. The origin of the Swiss name evidently differs from that of the Mauresque and Saracenic word.

[1] "Agasse, vieux mot qui vient de l'italien *gazza*, signifiant Pie."

That the Agassiz were descendants of French Huguenots, and were obliged to leave France at the revocation of the "édit de Nantes," is a tradition without any solid basis of fact to rest upon. Indeed, the name Agassiz existed as far back as the thirteenth century in the Canton de Vaud; but it is impossible to trace the family, because all the papers belonging to the Agassiz were destroyed in a fire at the parsonage of Constantine, in the Canton de Vaud, where the grandfather of Louis was settled as pastor, — a profession followed in the family for five generations.

Very likely an Agassiz married a French Huguenot; for at the time of the revocation the French Protestant exiles flocked into Switzerland, and settled in large numbers at Orbe and in the environs; almost completely filling the villages now known as Ballaigues, Vallorbe, and La Vallée de Joux, and it is possible that an Agassiz married among them; which may account for the tradition.

There is much to favour the belief in a connection of the Agassiz with some French family of the Cevennes, or of Provence; for the extraordinary imagination of Louis Agassiz points to a close connection with the children of sunny Provence, so well portrayed by Alphonse Daudet in his series of romances on Tartarin of Tarascon.

The family features are, however, entirely Swiss, and even Jurassic. In general, they are broad shouldered, thickly built, bony, with fair-coloured faces, and rather slow in their movements, — a type very frequently met all along the foot of the mountains of the Jura, more especially from Bienne down to Orbe and "la perte du Rhône."

The father of Louis was called Rodolphe Benjamin Louis; he was born March 3, 1776, and died Sept. 6, 1837. His mother, Rose Mayor, was born July 11, 1783, and died Nov. 11, 1867. Jean Louis Rodolphe Agassiz was born May 28, 1807, at the parsonage of Motier-en-Vuly, on the Lake of Morat, Canton of Fribourg. He was the fifth child, but the four others having died in their infancy, Louis Agassiz was the eldest child. As is the custom among all French-speaking people, he never used his full Christian name, but signed himself simply Louis Agassiz.

At the beginning of this century, and before coming to the parish of Motier, his father had been pastor at St. Imier, then a very poor and remote valley, lost among the mountains of the Jura. The loss of his children, one after another, and the great isolation of St. Imier, far away from his kindred and friends, led him to look for a better parish, and, in 1806, he came to Motier,[1] first as a "suffragan" (assistant) of the titular minister, J. R. Martin; afterward, on Aug. 31, 1810, he was elected "pasteur" (minister).

The parish of Motier consists of four small villages, located at the eastern foot of Mont Vuly, on the Lake of Morat, and containing in all five hundred inhabitants. Singularly enough, Vuly belongs to the Roman Catho-

[1] The name is sometimes spelled Môtiers, and to distinguish it from another Motiers in the Val Travers, it is called Motier-en-Vully. The spelling varies as well for Vully as for Motier, Motiers, or Moutiers, and is written sometimes Vuilly, Vuly, or Vully.

lic Canton of Fribourg; but in consequence of its situation on the extreme frontier of the Canton of Berne, it was invested with the right of com-burghership, "combourgeoisie," with Berne.

When, in 1530, the celebrated reformer, Farel, succeeded in converting the parish of Motier, the council of Fribourg complained to Berne of his preaching in the Vuly; deputies were sent therefore from Berne to meet at Morat together with four delegates from the four principal villages of the Vuly, who concluded to put the matter to vote under the direction of "Messieurs de Berne."[1] The reform movement received the majority of votes in the four villages of the Motier parish, and it has ever since been Protestant, notwithstanding the fact that it belongs to a very strong and uncompromising Catholic canton.

The Vuly is situated at the extreme end of a promontory, surrounded by water on three sides; on the east by the Lake of Morat, on the north by the river La Broye, and on the west by the Lake of Neuchâtel. The *Seeland* of Berne, comprised between Kersers, Treiten, Aarberg, and Bienne, constitutes, with the lakes of Morat, Neuchâtel, and Bienne, a very extensive sheet of water.

As soon as the young couple had left the trying climate of St. Imier, with its long and very cold winter, and had settled in the fine agricultural district of the Vuly, with its vineyards and orchards, prosperity and happiness greeted them from every direction. Four

[1] A name formerly used among the Swiss who speak the French language to designate all those in authority in the Canton of Berne.

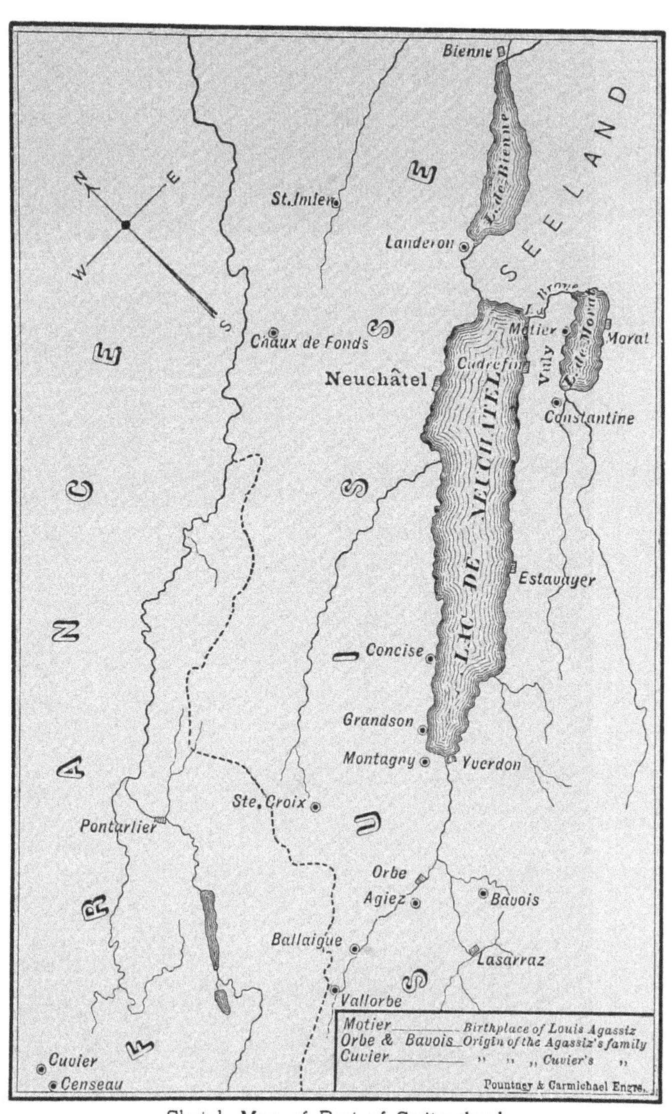

Sketch Map of Part of Switzerland.

healthy children — two sons, Louis and Auguste, and two daughters, Olympe and Cecile — were born to them; and although the parish was small and consequently of limited means, it was most gratifying to find themselves among relatives and friends; for pastor Agassiz had resided for some time at Constantine, a village near Avranches, where his father was minister, and it was during his stay there that he became acquainted with Miss Rose Mayor, his future wife, who was a daughter of the country physician at Cudrefin, a village only a few miles distant from Motier. It may be said that the inhabitants of the whole peninsula of Vuly, Cudrefin, and Constantine, greeted pastor Agassiz and his wife, as their own people returned.

Born and educated in such a place as Motier, surrounded by water and marshes, with the Oberland always in full view in front, and the summit of the Jura in the rear, it is no wonder that Agassiz became an ichthyologist and a glacialist. Everything which met his eye, from infancy until manhood, seems to have awaked in him a curiosity to know his surroundings. It was as natural for him to take to the study of fishes and of glaciers as it is for sons of seamen to go to sea, or for "vignerons" (vine-dressers) to go to the vineyard, or for the "gauchos" to ride on the prairies of South America, or for the Arabs to cross the desert on camels. It might almost be said that Louis Agassiz, as we shall see more fully by and by, was a remarkable instance of atavism of the Swiss lake-dwellers of prehistoric time.

Almost as soon as he was able to move alone, he

took naturally to water, like a young duck. All the fishermen became at once very fond of the little fellow, and there was a friendly rivalry among them to get him into their boats and show him how to catch fish. In a relatively short time he became a great favourite, and every one wanted to show the parson's son those neighbourly attentions which are of daily occurrence, and form a part, and an important part, of life, among all the country people residing in such isolated places as the Vuly.

A part of the duty of a minister in Switzerland is to look after the schools and even to take a part, and often not a small one, in the teaching. Parson Agassiz was a very successful and excellent teacher; indeed, in all his parishes, both at St. Imier and at Motier, and afterward at Orbe and Concise, his reputation as a teacher was far superior to his reputation as a preacher.

Louis was by far the best pupil of his father; for not only did he learn from him the elements, and lay an excellent foundation for his future education, but he caught from him his method of teaching, which was based entirely on the interest he always tried to awaken among his pupils in the subject of study. There is no doubt this was a family inheritance, and that it developed and attained its maximum with Louis. It may be said that Louis Agassiz was born with a true passion for teaching, as truly as that he was born a naturalist. As we shall see, he remained a teacher until the end of his life, changing his subjects of studies quite often, and showing a rather capricious character in many ways, except in his unalterable love of teaching.

Next to his passion for teaching, but developed before it, was his passion for collecting all sorts of objects belonging to natural history. As soon as he was able to catch fish, he brought them alive and placed them in a great stone basin of the fountain of the parsonage.[1] It is the custom in the Canton de Vaud and the neighbouring Swiss cantons to use boulders for basins, either to receive the water flowing from springs, or to hold the fruit of the vintage when the grapes are brought from the vineyard to be pressed and converted into wine. These boulders are generally of Alpine granite, and are cut into the proper shape, great care being taken not to break them, but to keep the block one great monolith. Such an Alpine boulder was the basin of the Motier parsonage, used as a *vivier* or aquarium by our young ichthyologist. It is not strange that, later in life, Agassiz became such an expert in boulders transported by glaciers; and it seems specially appropriate that one of them, transported from the Alps, should be his tombstone in America.

At the age of ten years, he was sent to the College of Bienne, to begin his classical studies; finishing them at the Academy of Lausanne in 1822–1824. He was a very clever student, but never had much inclination for mathematics and the physical and chemical sciences. He always showed great capacity for languages, becoming quite proficient in Latin and Greek, and when at Bienne, learning German and Italian, especially the

[1] "Le fils d'un pasteur du Canton de Vaud, dont le père, ne savait que faire *de ce garçon courant toujours les champs et toutes les rives du Lac à chercher des bêtes,*" as one of the contemporaries of his youth said.

former, which he spoke like a native. Geography was another favourite study.

His brother Auguste, younger by two years, joined him, a year later, at the College of Bienne, and the two brothers kept together during all their classical studies. The distance between Motier and Bienne—about seven leagues—was always made on foot; and, considering their extreme youth, it was quite a journey for such little fellows. But they were excellent walkers; and though at first the route was difficult, further on they crossed over the "Seeland," or marshy country, and the western shore of the Lake of Bienne, and reached their destination in eight hours, not over-fatigued. I heard both say, many years after, that it was shorter and easier to go home from Bienne than it was to go to the college at Bienne from home. Going home to pass the vacations was, of course, such an attraction that the roads seemed neither so long nor so dirty; besides, these journeys always happened at a fine time of the year; for college vacations then in all the Jura region were during the fall, from the first of September to the first of November. Louis's great strength and already vigorous vitality are finely illustrated by two anecdotes: Arriving quite unexpected from Bienne with his brother, he learned at Motier that his sisters with their cousins Mayor were at their grandfather's house at Cudrefin. At once he started again; but on his arrival at Cudrefin, he found all the young ladies enjoying a bath in the lake, at some distance from the shore. In order not to lose a moment, he jumped into the water with his clothes on, that he might the more promptly greet and kiss sisters and cousins,—the latter especially, for with one of them he was already in love; and to express his great admiration for this cousin he could find nothing better to do than to have her name tattooed on his left arm with sulphuric acid. The result was a severe attack of fever, and two or three weeks of forced inaction and rest for his arm, which betrayed his act to the whole family.

The two brothers divided their vacations between Motier and Cudrefin, where they enjoyed all the freedom of country life and the comforts of home and family surroundings. Their time was passed in fishing, rambling about with their grandfather, Dr. Mayor, in his daily excursions as a country doctor, in gathering in the second crop of hay, called "regain," and in the "vendanges." This last occupation was the most pleasing; for when the vineyards were richly laden with grapes, the labourers of the Canton of Berne—where there are no vineyards on account of the severity of the climate—gathered by troops from all around the shores of the Lake of Neuchâtel, to offer themselves as "vendangeurs," or vintagers, during the time of the gathering of the grapes and the process of crushing the juice under the "pressoir." It was a true festival, when everybody, young and old, was happy, singing throughout the day and calling from vineyard to vineyard. In the evening at supper, all the "vendangeurs" sat at the table with the proprietors, and the most proficient singers were called, one after another, to sing popular songs, which were closed in chorus by the whole company. Light wines were used very freely and the cups were continually filled and emptied during these very joyous and friendly repasts. For it must be kept in mind that the same labourers used to come every year, and the meetings were more or less reunions of old acquaintances, always full of reminiscences of preceding years.

Mingling so freely with the people round him, it would seem that Louis must have become an expert in horsemanship and the use of firearms, like almost all young men raised in country places in Switzerland. But this was not the case. Trusting to his great powers of swimming, and led on by his love for fish-catching, Louis was always ready to embark on skiff or raft of any sort. But on land it was very different. In Switzerland no one ever saw him on horseback; and it is doubtful if he ever tried to ride, except once in Brazil. All his life he invariably declined to mount horse or mule. As to shooting, he never possessed firearms, and never joined a shooting club, and probably never fired a single shot during his life in Europe. He is a very remarkable exception among the Swiss of the present century, for he never performed any military service. The great "Tirs Fédéraux" never attracted him, and his passion for natural history never carried him so far as to shoot birds or animals of any sort.

Curiously enough, when a student at the universities of Zürich, Heidelberg, and Munich, he became a great fencer; and one of his contemporaries and friends writes me that Agassiz was an excellent swordsman, using the rapier with great dexterity, and very ready to make use of it when discussions degenerated into quarrels and obliged the contestants to be called out. At Munich especially he was well known on account of several duels, which he fought to his credit and always successfully.

At the age of fifteen, it was decided that he should enter, as a clerk, the commercial house of his uncle, François Mayor, at Neuchâtel. But on account of his ability as a scholar, he was allowed to finish his college studies at Lausanne, where he studied "belles lettres" and the humanities for two years, 1822–1824. It was during his stay at Lausanne that he gave the first indication of the possession of the characteristics which afterward developed so strongly in him as a savant and as a man. Another of his uncles, Dr. Mathias Mayor, had great reputation as a physician, and a large practice at Lausanne and in all Romande Switzerland; and from him Louis learned the elements of anatomy. At the same time, he had his first opportunity to see a small museum of natural history, the Cantonal Museum, directed by Professor D. A. Chavannes, an entomologist of some note. The Helvetic Society of Natural History, founded at Geneva in 1816, had already collected a small band of Swiss naturalists of great talent, and Agassiz, during his stay at Lausanne, had the good fortune to meet Jean de Charpentier, director of the salt mines of Bex. He was much impressed by the fine face and splendid intellect of de Charpentier; a rare and perfect type of savant; and, influenced by the impression he received from his uncle, Dr. Mayor, Jean de Charpentier, and Professors Lardy and Chavannes, he resolved to become a naturalist; and, as the only way of accomplishing this, he asked his father to let him study medicine, his ambition at that time being to become a country doctor, with full time and opportunity to study the natural history of the Canton de Vaud.

Agassiz was a rather precocious young man; and coming to the age of manhood, he was at once a great admirer of the fair sex. It had always been a characteristic of the Agassiz family, well known among their kin of the Canton de Vaud; and Mrs. Rose Agassiz, who knew the family better than any one, and was always of a practical and far-seeing turn of mind, as far back as January, 1828, in one of her letters to Louis, says: "The sooner you have finished your studies, the sooner you can put up your tent, catch your blue butterfly, and metamorphose her into a loving housewife."[1] The unusual case of the old pastor Rudolphe, the grandfather of Louis, who married a second time at the ripe age of sixty-six years, was much commented upon at the time in the commonwealth.

In 1824 Louis went to the University of Zürich to pursue the medical course and prepare to become a doctor. Zürich was an old centre of culture, and he found there genial surroundings. The professor of natural history, Schinz, at once saw the rare qualifications of Agassiz to become a naturalist, and gave him every opportunity for the study of ornithology, which was his favourite branch. At Lausanne, although a very brilliant student, he was not so much above his

[1] "Louis Agassiz," by Mrs. E. C. Agassiz, Vol. I., p. 62.

comrades, for the Vaudois youth are almost all very bright and remarkably intelligent; but at Zürich it was different. The students there were, for the most part, German Swiss, rather slow, quiet, steady in their work, and not apt to awaken interest of any sort. Agassiz at once attracted the attention of all the students by his quick perception, his witty remarks, and more especially by his constant show of varied knowledge. One of them, many years after, said to me, "Agassiz knew everything, and he was always ready to demonstrate and speak on any subject. If it was a subject which he was not familiar with, he would study and rapidly master it; and on the next occasion, he would speak in such brilliant terms and with such profound erudition that he was a source of constant wonder to all of us."

At Zürich, Agassiz not only enjoyed the teaching of the zoölogist Schinz, but he also often saw the geologist Gesner Escher, the celebrated engineer of the Linth River, and became a great friend of his son, Arnold Escher von der Linth; a friendship which lasted all their lives, notwithstanding great difference of character and even of opinion, Arnold Escher being an observer in the field, while Agassiz was more of a laboratory, museum, and lecture-room man.

Not only did Louis Agassiz astonish his fellow-students at Zürich by the variety of his knowledge, but he was also a wonder to them in his capacity as a pleasure-seeker. At "Kommers" he was always first to come and last to go, his strong constitution requiring an absorption of food and drink which left all the others far behind him, and won for him the reputation of being what the French call a "belle fourchette," or one who wields a good knife and fork. From this time he took for his motto, "First at work, and first at play," and carried it through all his life, with rare interruptions. Such an ambition required a Herculean constitution, which he certainly possessed; but notwithstanding his strength, this burning of the candle at both ends no doubt finally shortened his life.

In 1826, after two years at Zürich, Agassiz went to the University of Heidelberg, where he stayed eighteen months. There he made acquaintances which influenced him as much as he could be influenced for the rest of his life. His studies took a more decided direction toward natural history, under the leadership of Professors Tiedemann (in comparative anatomy), Leuckart (in zoölogy), Bischoff (in botany), and H. G. Bronn (in geology and palæontology). While attending the lectures of Tiedemann and Bischoff, Agassiz became acquainted with Alexander Braun and Karl Schimper, two very brilliant botanical students; and they very soon became congenial and inseparable companions, not only during their courses at Heidelberg and afterward at Munich, but even during the first decade after leaving the universities.

Alexander Braun was the son of the postmaster-general of the Grand Duchy of Baden, who resided at Carlsruhe. As Heidelberg was very near Carlsruhe, Braun took home with him his three friends, — Agassiz, Schimper, and Imhoff (afterwards a distinguished entomologist of Bâle), — to pass their vacations. It was a great time, not only for the four intimate young naturalists, but also for the inmates of the house of Postmaster-general Braun. Besides a family of four children, two sons and two daughters, there was living in the family a young Swiss student, Arnold Guyot of Neuchâtel, then learning the German language, and preparing for the ministry. Postmaster-general Braun was very scientific in his tastes, and possessed one of the best collections of minerals then existing in Germany. His house, near one of the city gates, was large enough to admit and lodge comfortably all this company of young people, who rambled through the forests and fields, ransacking every corner where plants and animals were to be found. They had special rooms devoted to dissections, true laboratories; and here they brought their specimens, and for hours together discussed and theorized on all kinds of natural-history subjects.

I shall, farther on, speak at length of Alexander Braun and his younger brother Max, and also of Karl Schimper; but for the present I need only say that the two daughters, Emmy and Cecilia, were both very attractive, and soon received marked attentions, which afterward changed into courtship, from two of the visitors, Louis Agassiz and Karl Schimper. The younger, Miss Cecilia, or "Cily," as she was familiarly called, possessed a remarkable talent for drawing, and would have become an artist of repute if she had devoted her life to the fine arts. She was very sensible, affectionate, and unaffected, and soon felt the influence of Agassiz's attractive personality. Her talent for drawing was constantly brought into demand by the specimens of natural history which they collected; and when, in the spring of 1827, Agassiz was brought back to Carlsruhe a convalescent, after a very severe and dangerous illness of typhoid fever, the care bestowed on him by the whole Braun family, and more especially by Miss Cecilia, resulted in an engagement of marriage. During his sickness at Heidelberg, Alexander Braun did not leave him until he took him to his father's house at Carlsruhe, as soon as his convalescence allowed his removal. Later he accompanied him to Orbe, in Switzerland, where Agassiz's parents were then living, and left him only when he saw him rapidly gaining health in the life-giving air of the "Jura Vaudois," the native country of the Agassiz family.

As soon as he had recovered sufficiently to walk about, Agassiz, who always was an excellent pedestrian, began to explore the environs of Orbe, as a naturalist, collecting plants, insects, and his dear fishes; for he never lost sight of fishes for any length of time. The Jura Vaudois, during the summer months, is one of the most beautiful countries to visit, and, received as he was with great pleasure, as a guest at the houses of the pastors of Ballaigues, Vallorbe, Beaulmes, Ste. Croix, and other places, he was enabled to make a thorough exploration of Mount Suchet, 1591 metres in height, of the Dent de Vaulion, 1486 metres, of the Aiguilles de Beaulmes, 1563 metres, and of the valley of the Orbe River. At this time he first felt that love for the Jura Mountains which lasted until the end of his life. It was during this stay at Orbe that he wrote his first essay in natural history, a catalogue of all the plants growing in the Jura Vaudois. I do not know that it was ever published. I was led to think that it was, when Agassiz, speaking many years after of the Suchet Mountain, which he remembered with the freshness of a young man who has just visited them, said to me, "Do you know, it was the first work I did in natural history — an entirely botanical one." I have been unable to discover the work, however, and no one of his Swiss family has ever heard of it. It is only just to say that he was much helped in his botanical explorations round Orbe by the "suffragan," or assistant pastor of his father, Marc Louis Fivaz, like Agassiz an enthusiast in natural history. Fivaz soon abandoned all his researches in natural history, however, although he held the professorship of botany at the Academy of Lausanne, and became a voluntary evangelist in the interior of the west of New York, where he lived, in Tioga County, until his death.

This summer of 1827, when he had just attained his twentieth year, gave Agassiz his first real opportunity to work as a naturalist; and, as was always his custom in later life, he induced those around him to help him in collecting specimens and in making drawings. Besides his friend Marc Fivaz, who accompanied him in all his excursions, he put his younger sister, Cecile, to work drawing fishes and butterflies. His first two artists of natural history specimens, therefore, were two *Ceciles*, — the first, Miss Cecilia Braun, and the second, Miss Cecile Agassiz, his own sister, and the sister of his friend Braun.

CHAPTER II.

1827–1831.

Journey from Carlsruhe to Munich — His Evolution from a French-Swiss to a German Student — His Duel at Heidelberg — University of Munich — Doctor of Philosophy — Martius's Proposal to publish Spix's Fishes from Brazil — Publication of His First Great Work on Natural History — Doctor of Medicine and Surgery — Beginning of His Researches on the "Poissons Fossiles" — His Method of publishing His Works — His Desire to be the First Naturalist of His Time — Visit to Vienna — Return Home with Artist Dinkel — Life at Concise — Christinat — Opportune Help for a Trip to Paris — His Journey "en Zig-zag" by Way of Carlsruhe.

In the autumn, Braun wrote him about changing their place of study from Heidelberg to Munich. The attraction there was, that according to Döllinger, the instruction in the natural sciences left nothing to be desired. Accordingly, Agassiz left Orbe toward the end of October, 1827, and joined Braun at Carlsruhe, where he was welcomed by the whole family of the postmaster-general of Baden, more especially by Miss Cecilia, who had taken such good care of him when convalescent during the previous journey, and who had already made an excellent portrait of him, in coloured pencil or pastel, during the vacation of 1826.[1]

[1] "Portrait of Louis Agassiz at the Age of Nineteen," as frontispiece of "Louis Agassiz, his Life and Correspondence," by E. C. Agassiz, Vol. I.

On their way on foot to Munich they stopped first at Stuttgart, where the Royal Museum, already quite prominent, attracted much of the attention and even admiration of Agassiz. A splendid North American buffalo, and a piece of the hide of a Siberian mammoth, with hairs still attached, excited his wonder and his imagination; for he thought that the mammoth's teeth showed that it was a carnivorous animal; but the question that interested him most was, how this animal could have wandered so far north, and in what manner he died, to be frozen thus, and remain intact, perhaps for countless ages. He little realized that it was reserved for him some day to give the explanation, and that he was to be the god-father of the "Ice-age."

At Esslingen, Agassiz and Braun stopped to visit two botanists, and make exchanges of specimens. As Agassiz had a good collection of dried plants from the Jura Vaudois, and Braun of the Palatinate, they presented themselves, each with a package of dried plants under his arm, and were well received; more especially by Professor von Hochstetter, the father of the afterward celebrated naturalist, Ferdinand von Hochstetter of the "Novara" expedition round the world, and first director of the great Natural History Museum of Vienna.

On the fourth of November, 1827, Agassiz and Braun alighted at Munich; and a few days afterward they were joined by the other member of the *trio* of friends, Karl Schimper. Until then Agassiz had remained a true Swiss; his stay at Heidelberg had not been long enough nor successful enough, on account of his health, to make any change in him. But as soon as he was settled at

Munich, a change began in his mind, his thought, and even his body; and when he left Munich, three years later, on the fourth of December, 1830, he was entirely a German; so much so, that his Swiss friends had some difficulty in getting acquainted with him again. It was his first transformation; several others succeeded, as we shall see by and by.

Munich was then the most celebrated university in Germany, counting among its professors such men as Oken, Martius, and Döllinger. It was at the house of Döllinger that the three friends found lodging; occupying rooms which soon became laboratories, lecture rooms, and the rendezvous of many of their classmates. Agassiz afterward had occasion to give a vivid description of their student life, in his paper, "Erwiederung auf Dr. Karl Schimper's Angriffe" (Neuchâtel, November, 1842, 4to). In those days friendship reigned. Almost everything was enjoyed in common; work, pleasure, journeys, pipes, beer, purses, clothes, ideas political and philosophical, or poetical, and even literary. In fact, it was a constant, enthusiastic, intellectual life, lived at high pressure, lacking in nothing; not even student duels, and escapades of a more riotous nature, after grand "Kommers."

Agassiz enjoyed among the students the reputation of being the best fencer in the various students' clubs. The reputation was gained in this way: When at Heidelberg, an insult to the Swiss clan (Burschenschaft) of which Agassiz was the president was considered so serious among the students, that a challenge was sent by Agassiz to the German club. At a meeting of the

German students, a choice was made of one of their best swordsmen to meet him. Agassiz, however, would not accept such conditions, but said proudly, "It is not with one of you that I want to fight, but with all, one after another." They marched to the chosen ground, and in a few minutes four German students had received sword cuts on their faces; then the others who were to follow began to think that the affair had gone far enough, and although invited by Agassiz to take position, they offered honourable peace, and made an apology. After that, Agassiz was always chosen as arbitrator and judge at all the fencing clubs of the universities of Heidelberg and Munich. He was so carried away by his pleasure in fencing, that one day, without remembering to put on masks, he and his future brother-in-law, Alexander Braun, fell in with rapiers in hand, and after a few exchanges of thrusts, Agassiz made a cut in the face of his dear friend. Years after, when a professor at Neuchâtel, he appeared at a public fencing exhibition given by a tall and powerful negro fencing master, with success and credit.

Schimper, who was the oldest of the three, and whose imagination was the keenest and most original, exerted a great influence over Agassiz and Braun. His discoveries in regard to the morphology of plants gave him great advantage over the two others, who had not yet done any original work. But Agassiz was not the man to be long overshadowed by any one. He wanted to occupy the first place everywhere. Happily he escaped the danger to which Schimper succumbed, and, with the help of Braun, whose mind was the best balanced of the three friends, Agassiz kept out of temptation, and expended his impetuous nature in solid and difficult work on fishes, living and fossil.

Strange to say, with an allowance of only $250 a year, Agassiz managed constantly to keep in his pay an artist, Dinkel, to draw fossil and living fishes; and, occasionally, a second artist, Mr. J. C. Weber, to draw the Spix fishes and pieces of anatomy. They formed a sort of fraternal association. As Agassiz said, "They were even poorer than I, and so we managed to get along together." Their fare was certainly very simple; bread, cheese, beer, and tobacco being the main articles. Imagine Agassiz, with his scanty allowance, providing for two artists, besides Karl Schimper, and his younger brother, William Schimper. To be sure, Alexander Braun helped much also. But if we suppose that Braun got $300 a year from his father, six young men between the ages of twenty and twenty-five had to live upon less than $600 a year, out of which also they had to pay for their studies at the university, and provide themselves with instruments, and books, and clothing. Agassiz got a little money from the "Brazilian Fishes" and some other writing, with which he purchased a microscope — a rather expensive instrument — and several books; and he received, as a gift from Professor Döllinger, a copy of the finely illustrated work on living fishes, by the great French ichthyologist, Rondelet, of Montpellier. The editor Cotta sent him also a certain number of expensive natural history books.

"I cannot review my Munich life without great gratitude," Agassiz says. He was there a most happy and successful young man, using all the scientific resources existing in that large and progressive city; drawing round him comrades of the University, and even professors; and receiving visits from naturalists of renown, including the great anatomist Meckel. His room in the house of Döllinger, being the largest, was used as lecture room, assembly room, laboratory, and museum. Some one was always coming or going; the half-dozen chairs were covered with books, piled one upon another, hardly one being left for use, and visitors were frequently obliged to remove books and put them on the floor; the bed, also, was used as a seat, and as a receptacle for specimens, drawings, and papers. According to Agassiz, the tobacco smoke was sometimes so thick it might have been cut with a knife.

Agassiz was the most prominent among the students. His acquaintance was courted by all. He was specially considered with much pride by all the Swiss students, and was welcome both in the rooms and yards of the University, and at the students' clubs, "Bierbrauerei," and fencing rooms. He was considered a most amiable companion, never losing his temper, always smiling and apparently contented and happy. It is no wonder that he remembered so vividly his student life at Munich, and was always grateful for it. Although at Munich he learned embryology from Döllinger, who gave him personal instructions in the use of the microscope, and followed the lectures of the great philosopher Schelling, as well as the fascinating teaching of Oken, with his *a priori* conceptions of the relations of the three kingdoms into which he divided all living beings; he was not instructed then either in palæontology or in geology, the two branches of science in which he became afterward preëminent. The only teacher he ever had in those two sciences was Professor H. G. Bronn at Heidelberg, a rather second-rate palæontologist, but a very industrious and prolific writer.

Although Agassiz came to Munich for the special purpose of taking the degree of doctor of medicine, his studies soon drifted from those of a medical student to those of a true naturalist. This change was not made without warnings from his father, who became alarmed by the rather large expenses incurred by his son,[1] and more so by his neglect of his medical studies. But Louis Agassiz was born a naturalist, and a naturalist he must be; and, notwithstanding all sorts of difficulties, with the help of his mother, who always favoured his desires, he carried through his scheme of seeking a professorship of natural history.

His first step, in regard to graduation, was to secure not a title of medical doctor and surgeon, but of doctor of philosophy, which he won, in the spring of 1829, at the University of Erlangen. The excuse for so doing was a

[1] Agassiz's good heart had already, when at Heidelberg, led him to help, pecuniarily, Karl Schimper, and as soon as established at Munich, he sent Schimper money to pay the expense of his journey, and invited him to join them in their lodging. Schimper not only came directly, but brought with him his brother William. Agassiz's income was henceforth, on this account, limited in proportion. It must now suffice for the maintenance of a friend; for as soon as Schimper arrived, Agassiz gave him the key of his chest in which was his money; and, during the three years of his stay at Munich, he, in fact, gave to Schimper the means to satisfy all his needs — a rare example of generosity.

clever one. Martius had proposed to him to publish the fishes of Brazil collected by Spix and himself during their explorations on the Amazon, from 1817 to 1821. Spix having died in 1826, before finishing the publication of the zoölogical part, Martius, who possessed excellent judgment and great insight into character, saw at once the ability of young Agassiz as a describer of species, and proposed to him, during the summer of 1828, to take the fishes. The offer was certainly most flattering to Agassiz, then in his twenty-first year, and before he had yet published anything to recommend him. Martius assumed all the expenses and, of course, all the profits; and Agassiz received as his share only a few copies of the book, with an atlas of beautifully coloured fishes. Before the book was issued, Martius told him that it was important that his name should have the title of doctor of philosophy attached to it, and that at the same time it would help him to get a professorship of natural history.

However, that title would not do as a substitute for his medical degree, and, bracing his courage, he worked hard, and prepared his theses with great success; for when he received his degree of doctor of medicine and surgery, the 3d of April, 1830, the dean said to him: "The faculty have been very much pleased with your answers; they congratulate themselves on being able to give the diploma to a young man who has already acquired so honourable a reputation." It was nine months after the publication of his great work on the fishes of Brazil, a folio with ninety plates, which had attracted the attention of all naturalists, more especially of George Cuvier, the greatest ichthyologist of his time, who was then engaged on his celebrated work "Histoire des Poissons." Agassiz's work, which was dedicated to Cuvier, is written in Latin, and possesses the qualities so prominent in all Agassiz's publications; namely, accurate descriptions of the species, and excellent and even beautiful figures. It is most creditable in every way, and it furnished a sound basis for Agassiz's reputation as an ichthyologist of the first rank, although he was only twenty-two years old.

This first success was much enjoyed by his family and his friends, and prompted him to undertake a task which was sure to place him in the front rank as a naturalist, "hors ligne." Soon after the vacation of 1829, which was spent at Heidelberg, Carlsruhe, and Orbe, the director of the museum at Munich offered to Agassiz every facility to work at the collection of fossil fishes, allowing him to carry the specimens to his room. As the director of the Strasbourg Museum, the mining engineer Voltz, and Professor Bronn of Heidelberg had made the same offer, Agassiz, seeing what a splendid work was laid before him, did not hesitate to undertake it, notwithstanding the great difficulties, both material and scientific.

A few words should be said in regard to his method of undertaking work without being sure beforehand of the means to carry it on successfully. For example, in the case of his "Poissons fossiles," as we shall see, he first tried a M. Cotta, a publisher of Stuttgart, and when the latter failed to come to a final agreement for want of knowledge as to his part of the expense to be incurred (for Agassiz never knew beforehand what his work would be, even approximately, as to quantity of text and plates), he found that he could rely upon no one, but must himself publish his rather expensive work. Martius's work on Brazil was aided by a large subscription from the purse of the king of Bavaria. The only country in which it was possible to find a publisher for a very expensive work on natural history was France; and even there publishers required a certain number of subscriptions from the government before accepting the charge. "The Mineral Conchology of Great Britain" had involved great expense, without proper return, and was anything but a success in a pecuniary way. Goldfuss's "Petrefacta Germania," then in course of publication, 1826–44, was supported only by the personal sacrifice of the author himself. Even Cuvier's great work on the "ossements fossiles" was not successful from a bookseller's point of view, and without the help of the French government it would have been impossible to publish it.

If Agassiz had been a business man, or a good manager, he might have succeeded in having his work on the fossil fishes published in Paris, with a sufficient subscription from the Secretary of Public Instruction to carry on the work, if not at a profit, at least without loss on his part; for Cuvier was then publishing his "Histoire Naturelle des Poissons" in that way; and if that great work, finished after the death of Cuvier by Valenciennes, was not a pecuniary success, it entailed at least no expense upon its two authors.

Agassiz always acted as if he were a very rich man;

and now, taking an excellent artist, Joseph Dinkel, into his service, he had him draw all the fossil fishes he could find in the different museums; and, full of hope and never thinking of the morrow, he began his "Poissons fossiles," trusting to good luck and his power of persuasion. He was not patient enough to wait for the proper moment. With him time was money, and he pushed forward without regard to consequences. He had such self-confidence that it is almost amusing to see him writing to his father from Munich, Feb. 14, 1829: "I wish it may be said of Louis Agassiz that he was the first naturalist of his time, a good citizen, and a good son, beloved by those who knew him. I feel within myself the strength of a whole generation to work toward this end, and I will reach it if the means are not wanting." Strange to say, he attained his aim; if not the first, he was certainly one of the first naturalists of the nineteenth century; he was a good citizen of Switzerland and afterward of the United States, a devoted son to his father and mother, and beloved, if not by all, certainly by a great many of those who knew him in Europe and in America. This intuition of his capacity and strength, this thought that he had concentrated in him the powers of all his ancestors to observe and work on natural history, is something almost wonderful in its *naïveté*. It is not strange that his father was often frightened for the future of such a prodigy, for such certainly Louis Agassiz was.

During his stay in Munich, he went home only once, spending there the two months of October and November, 1829. His time was passed mainly at the parsonage of Orbe, and at Cudrefin, near his good grandfather, Dr. Mayor, who died during Louis's visit. As soon as he returned to Munich in December, 1829, he began his great work on the "Poissons fossiles," pursuing at the same time his medical studies, and, as we have seen, successfully passing his examinations, the subjects being anatomical, pathological, surgical, obstetrical, with inquiries into "materia medica," "medica forensis," and the relation of botany to these topics, as it is printed in his "Einladung."

As soon as his examination and the ceremony of receiving the degree from the rector of the University were over, he started for Vienna, where he passed almost two months. It was a great gratification to him to find that his reputation had reached Vienna, for he was received there, by professors and curators of museums, as "an associate already known." He looked specially at fossil fishes, and made memoranda of all the specimens, to be used afterward in his great work. His memory was so good, his eye so accurate, that many years afterward, when looking at fossil fishes at Neuchâtel, he one day said: "I have seen before another specimen of this same species in the museum at Vienna"; even going so far as to indicate the drawer in which it was stored. The director at Vienna, on being written in regard to it, answered that he had found the fossil fish where Agassiz had indicated that it was, and sent it at once. It was, as Agassiz said, of the same species.

Agassiz's last letter from Munich to his parents is dated Nov. 26, 1830. He left Munich the 4th of December, 1830, after settling all his accounts and expenditures, taking with him his draughtsman, Dinkel, to the great amazement of his father, who did not relish the arrival of his son's friend at his new parsonage at Concise, whither he had just removed from Orbe. His protest, that there was no room for another person, was of no avail. Louis wrote him that Dinkel was not in his pay, but was provided for by his publisher, M. Cotta, and that an agreement had been made for him to accompany Agassiz in future wherever he should go. Accordingly, one morning in December, 1830, he arrived with Dinkel, who was lodged in the neighbourhood, and came every day to Louis's room to draw fishes.

The parsonage of the village of Concise is beautifully situated close to the north of the church, with a terrace and garden overlooking the Lake of Neuchâtel, and commanding some of the most beautiful and extensive scenery in Switzerland. The room occupied by Louis Agassiz was on the first story, according to the way of counting stories in Switzerland; the second story, according to the American way, at the southeast corner of the parsonage, close by the churchyard. There he was seen constantly at the window, with his long moustaches, smoking, and hard at work with specimens.

The arrival of the parson's son, with his draughtsman, both dressed more or less as German students, with small caps on their heads, and long pipes in their mouths, greatly excited the curiosity of the quiet villagers. On Sunday they used to row on the lake, and, with long poles, passed their time in breaking the pottery jars, easily seen at the bottom through the transparent blue water, an amusement the young men of Concise were somewhat addicted to. How little Agassiz then thought that he was doing the work of a true barbarian, destroying pottery utensils which had belonged to his ancestors, the lake-dwellers of prehistoric time! After the discoveries of Keller at the Zürich Lake, Agassiz, remembering what he did in 1831 on the Lake of Neuchâtel, exclaimed: "How foolish I was! Dinkel and I have in sport broken dozens of important prehistoric pieces of pottery."

Almost a year of good work was passed at home, with nothing to disturb him, writing his "Poissons fossiles" and directing Dinkel's drawings. It was a great change after his rather boisterous student life at Munich. His habitual audience of fifteen or twenty persons, meeting daily in his room, and called the "Little Academy" by common consent, by both students and professors, was now reduced to Dinkel and his own family circle, with a few friends, relatives, and old acquaintances, who came in now and then, and were rather surprised, but unable to appreciate the work in which Louis was so deeply engaged.

Finally, the attraction of Paris proved too great, and to Paris he resolved to go, — a determination which he found not easy to carry into effect.

He had exhausted the paternal purse, and money was difficult to secure. At this critical moment came assistance, which was prompted entirely by friendly admiration and confidence in his future. An old friend of his father, a pastor of the Canton de Vaud, M.

Christinat, who had always been very fond of Louis since his childhood, came one day and simply put in his hand a sum of money sufficient for a journey to Paris. Helped also by his uncle Mayor of Neuchâtel, and his publisher Cotta, he was able to start on his much-desired journey to see Cuvier and enlarge his field of study of "poissons fossiles" in the great collections of Paris. Agassiz never forgot the generosity of Christinat, and after the death of his father he always considered Christinat a second father, feeling for him a true filial love.

He took "le chemin des écoliers" for Paris, passing by Stuttgart, Carlsruhe, Heidelberg, and Strasbourg, "to collect," as he says, "for my fossil fishes all the material I still desired, and to extend my knowledge of geology sufficiently to join, without embarrassment at least, in conversation upon the more recent researches in that department." To be sure, he took with him his draughtsman Dinkel; but the "zigzag" journey was made mainly in order to see the Braun family, and more especially Miss Cecilia and his dear classmate and friend, Alexander Braun, the best companion of his student life. Braun was the most reasonable by far of the trio, Agassiz, Braun, and Schimper; he was also the most steady and persevering as a student. His knowledge of geology was far superior to the knowledge of Agassiz in that branch of natural history; and Agassiz's avowed desire, as it were, to interview Alexander Braun on recent researches shows the method constantly used afterward by him for learning and keeping himself informed as to the progress and condition of geology. During all his stay in America, I was interviewed on my return from trips to Europe, or the interior part of the United States and Canada. Agassiz, with his remarkable memory, his keen perceptions of new discoveries, and his easy way of marshalling facts and using them afterward in his lectures or papers, would ply me with questions during two or three hours, in regard to all I had learned during my absence from Cambridge. It was a peculiar and rather original way of learning the more recent researches and the history of the progress of geology. But to make use of the facts without too much blundering required the splendid and rare qualities of an Agassiz.

In the letter to his mother, dated Carlsruhe, November, 1831, in which he speaks of his visit to the Brauns, he says: "I have added to my work on the fossil fishes one hundred and seventy-one pages of manuscript in French, written between my excursions and in the midst of other occupations." It is to be regretted that, with such facility as a scientific writer, he gave up almost all writing after 1837, trusting to secretaries, assistants, and stenographers.

CHAPTER III.

1831–1832.

FIRST VISIT TO PARIS — HIS RELATIONS WITH CUVIER — HUMBOLDT CHARMED WITH HIM — HIS VISIT TO THE SEASHORE AT DIEPPE — DEATH OF CUVIER — SKETCH OF CUVIER'S LIFE — CUVIER AND GEOFFROY ST. HILAIRE — THEIR DISCUSSION BEFORE THE FRENCH ACADEMY OF SCIENCE — CUVIER'S INFLUENCE ON AGASSIZ — DIFFICULTY OF GETTING AN OFFICIAL POSITION IN PARIS — APPOINTED PROFESSOR AT THE LYCEUM OF NEUCHÂTEL.

ON the 16th of December, 1831, after travelling by diligence for two days and three nights, on the road between Strasbourg and Paris, Agassiz and Dinkel alighted in the great crowded courtyard of the "Messageries Royales," rue Montmartre, so tired that they could hardly move hand or foot. The fatigue of these long journeys, in diligences, can hardly be realized now.

Packed in the rotunda with six often disagreeable neighbours, all breathing the same foul air, and jostled and even severely shaken from the bad roads, over which the diligence had to run, it was a great relief, first when the paved roads were reached, eighty miles before the arrival in Paris, and later when the great stagecoach finally turned into the courtyard of the "Messageries." It was an interesting sight for Agassiz to watch the diligences, arriving or starting, with prominent names, such as Forback, Bruxelles, Dunkerque, Calais, le Havre, Cherbourg, Brest, Nantes, Bordeaux, Bayonne, Toulouse, Perpignan, Montpellier, Marseilles, Lyon, Besançon, etc. Nearly every tongue was heard there, and the weary look of the arriving passengers was something not to be forgotten. He had not seen anything approaching the scene in Southern Germany, or even in Vienna. However, he soon found his way to the rue Copau, on the other side of the Seine, and alighted finally at the "Hôtel du Jardin du Roi," No. 4, just opposite the "Hospital de la Pitié," and close by the Jardin des Plantes. This third-rate hotel has always been a place of resort for naturalists, — French as well as foreign, — on account of its proximity to the great Museum of Natural History. The prices there were moderate and the fare good; a part of the old hotel, and the most desirable, was situated between a paved yard and a garden. It was here that Agassiz and Dinkel got a room. Sixteen years later, I saw Agassiz occupying the same room, quite proud to show that nothing had been changed in the arrangement of the furniture. There were the same shelves for books, where he placed, as he told me, the first works offered to him by both Cuvier and Humboldt; the only alteration being the removal of the small bedstead on which Dinkel slept.

As soon as Cuvier heard of his arrival, he sent for him; and Agassiz passed his second evening in Paris at the house of the great French naturalist. His reception was cordial and friendly, although with some reserve; for Cuvier was not a man of many words. His politeness was dignified, his manner that of a courtier accustomed to move among men in high office or of great social position. But he was kindly disposed and good hearted, his most notable characteristic being a kindliness directed right to the point, in order that no time might be lost. He always acted as if every hour was extremely valuable to him, working methodically; each day, each hour, having its task appointed in advance, and he was reluctant to be interrupted or interfered with in the course of his researches or thoughts. The first meeting was more than satisfactory to Agassiz, who had not expected such a friendly reception, which, in his own words, "more than fulfilled his expectations." He was absolutely astounded by the great erudition, the prodigious memory, and the extreme facility of Cuvier in passing from one arduous subject to another. Agassiz was more than charmed; he was actually astonished by the immense amount of knowledge accumulated in the brain of a single man. This first impression was never changed; and the more he knew of Cuvier, the more he admired him. Agassiz had found his master, and his leader for life.

After a few days of intercourse, Cuvier was so satisfied with the author of the Brazilian fishes that he gave him and his artist a corner in one of his laboratories, — the one devoted to fishes; for Cuvier possessed a laboratory for each of his works, where was accumulated everything pertaining to the subject, such as specimens, drawings, and books, in order not to lose time. The arrangement was a matter of wonder to Agassiz, who tried afterward — always in vain — to secure the same orderliness for himself. But later I shall give the reason why Cuvier was able to organize his time and laboratories, and why Agassiz always failed to do so.

Agassiz's main object in coming to Paris was to collect material for his "Poissons fossiles," and at the same time to become acquainted with Cuvier and some other French naturalists, in the hope that he might find an official position in such a great establishment as the Jardin des Plantes. Cuvier watched him attentively, and was so pleased with what he saw, that he deliberately decided to renounce his project of publishing, as he had intended, a work on fossil fishes; and with great generosity — too rare among savants — he placed at the disposal of young Agassiz all the materials, drawings, and notes which he had collected at the British Museum, at Avignon, and elsewhere, during more than fifteen years.

It was at one of his weekly Saturday evening receptions that Cuvier delivered into Agassiz's hands the drawings and notes, filling a large portfolio, brought from the study by his faithful assistant Laurillard. It filled Agassiz with greater joy than he had ever felt before or than he ever felt again, as he said many years after. Certainly such an acknowledgment from the greatest naturalist then living was a most gratifying and unexpected reward for all Agassiz's studies and efforts. It was most encouraging and auspicious for his future. All that Agassiz had expected, and even this was with grave doubts, was that, perhaps, Cuvier might be induced to allow him to assist in finishing the work with him, just as he had lately allowed Valenciennes to help him to finish his "Histoire naturelle des Poissons vivants." Of course the gift of Cuvier was highly appreciated by his parents on the shore of the Lake of Neuchâtel. It was most gratifying for them to see their dear Louis so well treated. But, alas! just at this time the small sum of money he had brought with him began to run very low; and there was no immediate prospect of replenishing his purse unless he accepted an offer from J. Daudebard de Férussac to take the editorship of the zoölogical part of his "Bulletin des annonces et des nouvelles scientifiques," which would yield an additional thousand francs a year, but would require two or three hours' work daily.

One of his first acquaintances at Paris was Alexander von Humboldt, then a star of the first magnitude among the numerous great French savants. If Cuvier's welcome was somewhat reserved and marked by formal politeness, lacking cordiality, Humboldt's reception took a form of indulgence and kindness which warmed Agassiz's heart. From their first meeting at the apartments of Humboldt, in his working room in the rue de la Harpe in the Latin Quarter, a mutual attraction was felt; and the terrible Humboldt, the fear of all savants and of all the great salons of Paris, took a fancy to the young Swiss naturalist. He took him to breakfast at his usual café-restaurant, the celebrated Café Procop, rue de l'Ancienne Comédie, near by; and there, as was his custom, hardly taking time to eat, he talked incessantly of his experience among the electric fishes in Venezuela. Agassiz, who was all attention, did not interrupt him once, — certainly a great mark of admiration on the part of Agassiz, who was himself a great talker, — and after three hours together, they separated at the door of the Mazarine Library, charmed with each other.

To have pleased a man so sarcastic as Humboldt was not a small triumph. He was conquered by the juvenile enthusiasm, the extraordinary optimism, of Agassiz. As one of the friends of Agassiz says, "Lui (de Humboldt) qui était si mordant de nature n'est qu'affectueux et plein de sollicitude en écrivant ou en causant avec son jeune ami."

In some way Humboldt learned through the publisher, Cotta of Stuttgart, the straitened circumstances under which Agassiz was labouring, and therefore simply enclosed in a letter a "billet de la Banque de France" for one thousand francs, begging him to accept it. Agassiz, in his letter of thanks, dated the 27th of March, 1832, calls Humboldt his "benefactor and friend," and confesses that his kind and helpful hand has unexpectedly rescued him from a distressing position, and that now he is again in hope of devoting his whole powers to science. Humboldt was then minister plenipotentiary of Prussia to the court of the Tuileries; and his timely help won for him an admiration which ended only with the last day of Agassiz's life. As Agassiz himself says, "I was pleased to remain a debtor of Humboldt, for I have never returned the sum he bestowed at such an opportune moment."

Agassiz-like, as soon as he was recovered from his despondency in regard to money, his buoyant spirit led him at once on a journey to the seashore of Normandy in company with Alexander Braun, who had joined him in Paris six weeks after his arrival, and Dinkel. They walked all the way from Havre to Dieppe, enjoying to the full the spectacle, so new to them, of living sea-animals, bringing back from that too short visit many new ideas, cheered and stimulated by "the great phenomena presented by the ocean in its vast expanse."

A few weeks after his return from Normandy, Agassiz sustained a great loss, — a loss which affected the rest of his life, — in the death of his master, George Cuvier. Since Carnival and during the whole spring cholera had been raging fearfully in Paris, greatly increasing the death rate; some quarters, however, like the Jardin des Plantes, had been almost free from the terrible scourge, but there it at last made its appearance, and one of its most illustrious victims was Cuvier. Sunday, the 6th of May, 1832, Agassiz, as was his custom, worked all the day until dinner time at five o'clock in Cuvier's study. During a conversation, Cuvier, seeing how intense Agassiz's application to work was, said to him: "Soyez prudent, et rappelez vous que trop de travail tue." On the next day Baron Cuvier, who, in 1831, had been created by King Louis Philippe a peer of France, when about to ascend the tribune in the Chamber of Peers, at the Palace of the Luxembourg, to deliver an address, suffered paralysis. He was carried home, and rallied, but died on Sunday, May 13, 1832, the immediate cause of his death being an attack of cholera.

The unexpected and somewhat premature death of

Cuvier at the age of sixty-three — for his life might have extended ten and perhaps fifteen years longer — had a very serious effect, which cannot be overestimated, on the future of Agassiz. Cuvier was the only man who exerted a scientific and personal influence over Agassiz; from him, and him alone, Agassiz would accept advice, and be guided in his work. He recognized in him his master, and the young charmer of Switzerland found in him another more powerful than himself, and especially more practical in his life and work. At first the formal politeness of Cuvier chilled him, and he says, "I would gladly go away were I not held fast by the wealth of material of which I can avail myself for instruction." But this first impression soon passed away, and an unbounded admiration replaced it.

Some details are necessary to understand the course taken by Agassiz, and the singular resolve to leave Paris, at that time the Mecca of all naturalists and savants, to settle as a professor, with a very small salary, in a small town of less than six thousand inhabitants, in a hybrid country, half Swiss, half Prussian, lost in Central Europe.

Cuvier, son of an officer of a Swiss regiment, in the French service, and nephew of a Protestant clergyman of talent, was called to Paris, after the revolution of the 9th Thermidor, by young Geoffroy, — celebrated since as Etienne Geoffroy Saint-Hilaire, — and attached to the Jardin des Plantes, as substitute for the professor of comparative anatomy. The reading of some manuscript papers on natural history, sent by Cuvier, had excited in the enthusiastic mind of Geoffroy such an admiration that he wrote him, then in Normandy, acting as tutor in a nobleman's family, "Come and play among us the part of Linnæus — of another legislator and ruler of natural history." This was at the beginning of 1795. Cuvier soon rose to the front rank, and even to so high a position that, after 1817, the year of the appearance of his "Règne animal," he was recognized by all European naturalists as unquestionably the leader. From that moment Cuvier developed a love of power and a tyrannical spirit which surprised and grieved some of his best friends. He became overbearing and impatient of any opposition to such a degree that in 1830, during the celebrated discussion before the French Academy of Science, occasioned by the publication of the "Principes de philosophie zoologique," in May, 1830, a rupture occurred with his life-long friend Geoffroy Saint-Hilaire. Cuvier and Geoffroy became irreconcilable antagonists, but remained personally friendly, though the intimacy which had existed between them during more than thirty years ceased, as much through the fault of Geoffroy as of Cuvier. In the discussion Geoffroy was very overbearing, and assumed a rôle which extremely irritated Cuvier. It is generally admitted now that Cuvier went too far, although he refuted, with a surprising number of facts, the arguments presented by Geoffroy on the six great problems: (1) The pre-existence in natural history of the genus; (2) the unity of organic composition; (3) the value of classification; (4) the fixity of species; (5) the final cause; and (6) the succession of organic life on earth. Thanks to his genius and his unrivalled talent of exposition, Cuvier won before the Academy; but it was plain that the general public was against him and in favour of Geoffroy. During the last two years of Cuvier's life the discussion was continued, not before the Academy, but in public lectures at the College de France. Cuvier, with renewed vigour, assailed the unity of organic composition and any general conception in natural history. As Isidore Geoffroy says: "Disciple, Cuvier ne pouvait l'être de personne, et par les tendances propres de son esprit, moins de Geoffroy Saint-Hilaire que de tout autre; il devint donc adversaire." We may say, to the credit of Geoffroy, that his admiration of Cuvier was not diminished, and at his tomb, with great emotion, and in words of sincerity which had their source in his heart, he proclaimed him "le Maître à tous!"

Agassiz felt strongly the influence of Cuvier; he had repeated occasion to see and compare Cuvier and Geoffroy, and the superiority of Cuvier was so undeniable, that many years afterward, when the question of fixity of species, descent, and succession of forms again arose with Darwin's "Origin of Species," he did not hesitate for one moment to oppose a doctrine so full of hypothesis and so contrary to the teachings of his master: "le Maître à tous!" Agassiz had promptly received the good will and protection of Cuvier, and it is most probable that, if Cuvier's life had been spared, he would have obtained, through his influence, a professorship or some place in Paris. For Agassiz was determined not to be a country physician, but to support himself as a naturalist. It is always very difficult in Paris to get a scientific position, on account of the great number of aspirants always waiting for a favouring opportunity. After a few months, Agassiz soon realized his superiority over all the young and even old naturalists, and acknowledged only one master, — George Cuvier! For Agassiz was not naturally self-distrustful; he knew his worth, and it was rather humiliating to him to be placed beneath certain savants whose merits and capacities were far below his. He had to reckon with those who held what is called in Paris "positions acquises," that is to say, with savants who had been gradually promoted from very modest places to higher positions. Cuvier's death left vacant a large number of places, and a regular scramble to occupy all the positions he had held began in earnest as soon as he was buried. Not being a Frenchman by birth, Agassiz was at a disadvantage; although an Englishman, Henri Milne-Edwards had the good fortune to push himself forward, and finally succeeded Cuvier. But Edwards, who was older than Agassiz by several years, had been educated in Paris, and knew how to make use of influence. He made loud claims to being a Frenchman born, because he was accidentally born in Bruges in Belgium during the occupation of that country by the armies of the French Republic. Another competitor was Valenciennes, also older than Agassiz, and an assistant of Cuvier, who had begun with Cuvier the publication of the "Histoire naturelle des Poissons vivants." To be sure, the publisher of the work proposed to Agassiz to join Valenciennes as a collaborator; but Humboldt, who had always taken a strong interest in Valenciennes' welfare, rather discouraged the association, knowing well that Agassiz would soon extinguish all Valenciennes' future prospects. Humboldt exerted a strong influence over Agassiz. As minister of Prussia in France, he cunningly worked to detach Agassiz from Paris, increasing rather than diminishing the obstacles and difficulties Agassiz found there, acting in accord with the Prussian governor at Neuchâtel, and M. Louis de Coulon, a rich and most benevolent Neuchâtelois, who wished in some way to attach Agassiz to the Lyceum of Neuchâtel. After the death of Cuvier, Agassiz, with his independent character, was discouraged and distressed by the constant intrigues going on under his eyes in restless Paris. On the other hand, he was still mindful of his happy days in Germany, and desired to return there as a professor in some German university.

Humboldt, little by little, persuaded Agassiz to accept a very modest — altogether too modest — position as professor at the Lyceum of Neuchâtel as a stepping-stone and a preliminary position to a professorship at Berlin or some other German university. Agassiz hesitated, for he knew very well that Neuchâtel was too small a place, and devoid of all resources in natural history; and his thought was at first to settle at Lausanne, or preferably at Geneva, then already a great scientific centre. But Humboldt and Coulon united their efforts, and at last secured the acceptance of Agassiz, who, in September, 1832, left Paris, to the great joy of all the young French naturalists of the capital; for he was a formidable rival taken out of the way.

Agassiz always disliked intrigue; he was frank and very earnest, and, although inclined to authority and adverse to divided power, he was too little French, or more correctly too little Parisian, in character, to like living in a society in which intrigue was as necessary as scientific knowledge to success. He had too high an opinion of science to make compromises and constantly bargain for position, influence, and honour.

With the death of Cuvier vanished all his hopes of a great journey beyond Europe, — a desire which had pursued him ever since he began the study of natural history at Zürich with Schink. What he heard in Paris of the great success of Victor Jacquemont in India, and of Alcide d'Orbigny in South America, had increased tenfold his wish to be a travelling naturalist, and the long account given to him by Humboldt of the equinoctial regions of the New World increased, if possible, his cherished determination to make an exploring journey. Cuvier had told him that after the return of Jacquemont and d'Orbigny, then daily expected, the annual appropriation at the disposal of the Museum would be in part free, and might be bestowed on him. His dreams of seeing the great Amazon River were revived, but were not destined to be realized till more than thirty years afterward, under other auspices, and under much more fortunate conditions.

If Agassiz had been able to make a great exploration into the interior of a continent, or around the world, when he was between twenty-five and thirty-five years of age, what a harvest of facts he would have brought back with him! It is much to be regretted, both for himself and for the progress of natural history, that he did not enjoy that privilege.

It is useless to express regret as we see him burying himself in such a remote place; for wherever Agassiz went he carried with him the torch of science, and obliged all the scientific world to look at him and give close attention to what he said and did.

CHAPTER IV.

1832–1835.

AGASSIZ'S FIRST ESTABLISHMENT AT NEUCHÂTEL—FOUNDATION OF THE "SOCIÉTÉ DES SCIENCES NATURELLES," ON THE 6TH OF DECEMBER, 1832—AN OFFER OF A CHAIR AT THE UNIVERSITY OF HEIDELBERG DECLINED—LETTER OF HUMBOLDT—ENGAGEMENT OF ALEXANDER BRAUN WITH MISS CECILE GUYOT AND THAT OF KARL SCHIMPER WITH MISS EMMY BRAUN BROKEN OFF—MARRIAGE OF AGASSIZ WITH MISS CECILE BRAUN—PUBLICATION OF THE FIRST PART OF THE "FOSSIL FISHES"—FIRST VISIT TO ENGLAND IN 1834—"MONOGRAPHIE DES ECHINODERMES"—DES MOULINS'S WORK ON THE SAME SUBJECT—CRITICISMS OF HUMBOLDT AND VON BUCH—SECOND VISIT TO ENGLAND IN 1835—BIRTH OF A SON—FOUR LETTERS TO PICTET AND NICOLET.

IN November, 1832, Agassiz was established at Neuchâtel as professor of natural history in a small college, at a salary of eighty louis (about $400), and with an appointment of only three years' duration. What tempted him greatly was the opportunity to live in Switzerland, near his family, complete independence in regard to his teaching, and a belief that, notwithstanding the small salary, the expenses of living were so much less than in large cities like Paris or Munich, that $400, or 2000 francs, would "keep him above actual embarrassment." Independence he got; and independence was a strong trait in his character, and one which explains several of his otherwise rather peculiar decisions at different periods of his life. But insufficiency of means, resulting from his want of business capacity, assailed him from the first moment of his arrival in Neuchâtel. As to its being a stepping-stone to a position at Berlin, that expectation was never realized; all prospects in that direction having been entirely barred, as we shall see, by the part he took in the glacial question five years later.

A college in a small town of five or six thousand inhabitants, like Neuchâtel in 1832, and after the political and very grave disturbances which occurred there in 1831, as a consequence of the French Revolution of July, 1830, was, of necessity, a very limited institution. The number of pupils, all told, was below one hundred; and there were absolutely no materials for study, no collections, not even a room to be used for the new class. Agassiz was obliged to deliver his lectures at the City Hall, in the room of the tribunal of the justice of the peace. With his impetuous and optimistic spirit and his impulsive nature, he went to work, and, without losing a minute, he undertook to form a centre of scientific culture with the rather scanty and rough material at his disposal. With the help of the two Louis de Coulons, father and son, — two of the most devoted, and, at the same time, most modest naturalists, — Agassiz arranged a provisional museum in the Orphans' Home, bringing there the already numerous specimens of natural history collected by himself in Germany, Switzerland, and France.

Less than a month after his arrival and the delivery of his inaugural lecture, "Upon the Relations between the Different Branches of Natural History," which was given on Nov. 12, before all the educated and intelligent men Neuchâtel could assemble, his father included, — on the 6th of December, 1832, he founded the "Société des Sciences Naturelles de Neuchâtel," in the parlour of M. Louis de Coulon, Sr., who was elected president, while Agassiz was made secretary. During the first six years of its existence the society met at M. de Coulon's private house. It was rather more a scientific club than a true society, meeting twice a month from November until May, and monthly only during the rest of the year. The annual subscription was moderately placed at three francs (sixty cents). Only six persons founded the society, — Agassiz, Auguste de Montmollin, the geologist, Louis de Coulon, Jr., and three others. At the end of the first three years of its existence, the number of fellows was only twenty-five, and not more than six or eight members were often present at the meetings. Agassiz was the leading spirit; and he wrote the proceedings of the sections of natural history and medical science for the years 1833, 1834, 1835, and 1836. His first contribution was a "General Report on the Progress of Natural History during the Last Few Years," in which he paid a tribute of admiration to George Cuvier, and at the same time lamented the recent death of "Ce héros de la science"; and declared that the only way to success is by "the conscientious observation of nature."

Not satisfied with delivering a course to his class at the college, he gathered round him a select and limited audience of persons desirous to hear him on zoölogy, botany, and the philosophy of nature. And when the weather permitted, he used to take all his pupils, young and old, on excursions into the field; visiting, among other places, the celebrated quarries of the Neocomian at Hauterive, the summit of the Chaumont Mountain, and the shores of the lake. It was a spectacle worth seeing. One of those who enjoyed these excursions said to me: "Agassiz was at his best, passing from a plant to a fossil; from physical geography to a fish, a snail, a bird, an insect, anything that came in his way; always ready to discourse for hours, and, as usual, full of all sorts of new schemes. Time passed only too quickly in his company."

Everything then seemed to smile on him; it was a sort of triumphal entry into life. A few days after his installation at Neuchâtel, he received, on the 4th of December, 1832, a proposal to present himself, if he wished, in place of Professor Leuckart, at the University of Heidelberg, which he declined to do. He consulted Humboldt about the call from Heidelberg, in a letter published by Mrs. Agassiz, Vol. I., pp. 213-217; but as Mrs. Agassiz was unable to give Humboldt's answer, I will give it in full in French, translated from the German by Agassiz himself for his uncle Mayor.

BERLIN, 29 décembre, 1832.

LETTRE D'ALEXANDRE DE HUMBOLDT À LOUIS AGASSIZ,—

Je n'ai point d'expression, mon cher Agassiz, pour vous témoigner quel grand plaisir me procure chaque ligne que je recois de vous, ainsi que les marques d'amitié que vous me donnez. Je ne puis excuser le retard que j'ai mis à repondre à votre dernière lettre que par mon bras manchot et la vie pénible que je mène entre Berlin et Potsdam, où je vais retourner dans un instant et passer quelque jours avec le roi. Vous m'aimez assez pour ne pas vous en fâcher. Ceci n'est pas non plus une lettre, mais seulement une marque de ma gratitude et mon opinion sur la proposition si honorable qui vous est faite d'une chaire de professeur à Heidelberg. Je reconnais, mon cher ami, le grand sacrifice que vous ferez en refusant, mais sans pouvoir de loin apprécier votre position à Neuchâtel, et surtout ce qu'elle peut avoir de fixe, je penche cependant pour vous engager à y rester. Ce pays est en quelque sorte votre patrie; on vous a reçu là, à ce qu'il me parait avec beaucoup d'empressement. Il est vrai que par ce choix, vous perdez sensiblement en argent, mais à Neuchâtel vous avez plus de temps à vous, et vous vivez dans une ville riche, où certainment, en considération du sacrifice que vous faites, on aidera peut-être plus qu'ailleurs, soit par reconnoissance, soit par un sentiment d'honneur ou même de vanité de vous posséder (le patriotisme prend toutes ces tournures) dans la publication de vos deux grands ouvrages qui doivent être le but essentiel de votre vie. Un homme de votre talent et de votre savoir, lorsqu'il aura publié ces deux ouvrages, sera placé si haut que de telles offres par des universités allemandes ne sauraient manquer d'être renouvelées. Il est vrai que je préférerais Heidelberg à toute autre, même à Berlin, où maintenant l'étude des sciences naturelles est assez négligée. Cependant je désirerais qu'il fut bien connu par les feuilles publiques que vous avez eu cet appel et que (sans menacer indélicatement) vous missiez à profit votre refus pour fixer votre position à Neuchâtel, pour accélérer l'achat de votre collection et pour obtenir la promesse de quelques souscriptions considérables pour vos ouvrages. Je ferai tout ce qui dépendra de moi auprès de M. Ancillon et suis bien certain de ne rencontrer ici aucun obstacle, mais je crains qu'à Neuchâtel même on ne soit un peu intimidé par une dépense de 600 louis, au moins à présent. En Allemagne, il est maintenant très difficile d'obtenir quelques cents louis d'un gouvernement pour quoique ce soit. On prétexte toujours les probabilités d'une guerre, à laquelle du reste personne ne croit. L'essentiel maintenant me paraît (et je vois avec plaisir que vous y visez constamment) que vous fassiez voir au monde, même avec des figures moins soignées, ce que vous avez si admirablement bien examiné, et si vous ne pouvez pas publier en même temps ces deux ouvrages, donnez la préférence aux fossiles. Vous le pouvez d'autant mieux que vous avez si heureusement découvert les rapports qui existent entre votre classification des poissons et la succession des formations géologiques. Je suis bien impatient de la connaître dans tous ses détails. Ne négligez pas de consulter sur les fossiles en général la traduction du manuscrit de de la Bèche par Dechen; c'est lui qui, à mon avis, rend le mieux compte de l'état actuel de la géologie. C'est aussi l'opinion de de Buch.

Si vous ne pouvez pas commencer promptement la publication de votre ouvrage, il faudrait nécessairement pour ne pas être volé, publier de suite sous vos yeux un mémoire en français dans lequel vous rendriez compte de vos idées générales sur la classification des poissons, leur distribution géographique et géologique, etc. Afin que votre ouvrage produise tout l'effet qu'il doit faire, restreignez vous aux poissons et ne donnez que quelques indications générales sur les autres organismes.

Je vous embrasse tendrement, mon cher Agassiz; assurez votre bonne mère de toute mon estime.

Votre A. HUMBOLDT.

This letter is, perhaps, the most important and affectionate one from Humboldt to his young friend Agassiz.

Thanks to Humboldt's personal application to the Prussian government, and the initiative subscriptions taken by Louis de Coulon at Neuchâtel, a sufficient sum of money was obtained to purchase, for the newly created Museum of Neuchâtel, the collections of Agassiz. He received the round sum of 600 louis, or almost $3000.

With his optimism, always ready to go beyond all reasonable bounds, he thought that in coming to Neuchâtel he had made his fortune; and his first desire was to get married. As is often the case with students, the trio, Agassiz, Braun, and Schimper, had promptly fallen in love. They were surrounded by too many young sisters and friends' sisters not to succumb. Agassiz's choice was Cecile Braun; Schimper became engaged to her older sister Emmy, and Braun himself was soon enamoured of a sister of Arnold Guyot, also called Cecile. As shown by the result, it would seem that it would have been well if the three engagements had been broken off. Alexander Braun, the most reasonable and practical of the three, had the good sense not to go too far. His regard for Cecile Guyot of Hauterive, near Neuchâtel, soon took the form of friendship. No public engagement was announced, prudence on both sides keeping the matter rather quiet, until, by common consent, Mademoiselle Cecile Guyot instead of a "mariage d'inclination," considered by her very practical family as a great "imprudence," agreed to a "mariage de convenance" at Neuchâtel, — a "parti fort avantageux," according to Arnold Guyot, — and, instead of becoming Madame Braun, was contented to call her old sweetheart "son bon frère Alex." [1]

Although Alexander Braun was a great admirer of the botanical genius of Karl Schimper, he soon saw the weak point of his character. After waiting several years from the time of the engagement in 1832, Miss Emmy Braun realized too well the unfitness of Schimper, and, with the help of her brother, broke the engage-

[1] See "Alexander Braun's Leben," by Mrs. C. Mettenius, where are many details of the whole affair, even including letters of Braun to Cecile Guyot.

ment in 1840. Miss Emmy Braun was born in January, 1807, and was as gifted as her younger sister Cecile, being an excellent musician. In the spring of 1841 she married Mr. Eichhorn, Hofmuziker at Carlsruhe. All four of the Braun children were talented.

As for the engagement of Agassiz and Miss Cecile Braun, although Alexander and his mother were not very enthusiastic in regard to it, — for both saw how fully Louis was engrossed in his studies and in himself, and realized his tendency to fly from one subject to another, and his want of steadiness and of business capacity, — nevertheless during the vacation of 1833, it became "un fait accompli"; and in October Agassiz brought his wife home to Neuchâtel, to a small apartment "au faubourg du Lac," No. 21.

Mrs. Cecile Agassiz, born at Carlsruhe the 29th of July, 1809, was a lady with regular and fine features, slender, and of very dark complexion, so much so that she looked more like an Italian or Spaniard than a German. She possessed rare artistic talents, being a pupil of an artist of some repute, Marie Ellenrieder of the Nazarean school, after the style of Fra Angelico. Before her marriage she made many aquarelles of fossil and fresh-water fishes for her "fiancé," remarkably exact and well executed, which rivalled those made by the painter Dinkel. She also painted specimens of plants for her brother Alexander. Besides this, she was well acquainted with German literature, and was generally an accomplished young lady. She was a great favourite in her family, and was widely acquainted in Carlsruhe. She was greatly disappointed in Neuchâtel; everything was so different from her delightful home at Carlsruhe. She did not speak French fluently, and possessing to a high degree the German placidity which borders on complete indifference, she was not well impressed by what she saw, and from the first disliked all Agassiz's friends and acquaintances. Accustomed to the beautiful green fields and forests of the vicinity of Carlsruhe, she found herself enclosed by dusty or muddy roads, by high vineyard walls, and the rather inhospitable aspect of the houses: all this, with the reserve and rather cold manners of the inhabitants, disposed to copy the formality of the Prussian court, displeased her so much that she soon greatly disliked the "Neuchâtelois," Neuchâtel, and even Switzerland. For her, Carlsruhe was paradise on earth, and her only wish was to return and live there.

Agassiz, during the first three years of his married life, showed more than at any other period the brilliancy of his rare intellect, the deepness of his devotion to the progress of natural history, and the greatness of the effort he was able to make to place himself among the foremost naturalists of the time.

During the spring of 1834, the first number of the "Poissons fossiles" came out, and made a great sensation among geologists and zoölogists. The subject had until then baffled all palæontologists, no one having ventured to go deeply into it, on account of osteologic difficulties and the material obstacle of drawings. The most difficult to please declared the work remarkably executed, and Agassiz received approbation and congratulations from every quarter. Undoubtedly the first "livraison" fully deserved such a reception. Agassiz never surpassed, perhaps never equalled, that first number of the "Fossil Fishes." It is the work of a great master.

A few days after, in May, 1834, another memoir, also very remarkable, was read before the Natural History Society of Neuchâtel on "Quelques Espèces de Cyprins du lac de Neuchâtel" ("Mémoires Soc. Sc. nat. de Neuchâtel," Vol. I., p. 33). In this Agassiz shows his tendency to create new genera and his admirable talent for description of species and for classification.

In August, 1834, Agassiz made a long-desired visit to England. Buckland, Lyell, and others received him with open arms. His visit coincided with François Arago's journey to collect material for his academic eulogy of Watts, and as he had become well acquainted with Arago during his sojourn at Paris in 1832, they were much together, meeting at Oxford at the hospitable home of Buckland, and travelling together to Edinburgh, and back to Paris.

Agassiz found such a wealth of fossil fishes that he wrote at once to his artist, Dinkel, to come over. One of the rooms of the Geological Society, then at the Somerset House, was generously placed at his disposal by the society, and as soon as he had collected there some two thousand specimens, he began in earnest his studies of comparison, determination, and classification, and directed Dinkel to draw all specimens worthy of being reproduced for his great monograph, or even such as might prove useful afterward for general description. Dinkel remained in England for several years, either in London, or at the country seat of Philip Egerton, near Chester, and at Enniskillen in Ireland, and made one of the best and most valuable collections of drawings of fossil fishes, which was afterward purchased by the subscriptions of English geologists, and presented to the British Museum.

On the 3d of December, 1834, at a meeting of the Neuchâtel Society of Natural History, Agassiz delivered a lecture on the present state of natural science in England, on the splendid collections of fossils and living animals there, and more particularly on the great progress and extraordinary enlargement of the Zoölogical Garden of London.

The echinoderms had already attracted much of his attention. The peculiar beauties of these fossils, their great numbers around Neuchâtel and in the Jura Mountains, and the ease of identifying them, even from fragments, led him to undertake a "Monographie des Echinodermes." At the meeting of the Neuchâtel Society, Jan. 10, 1834, he made a communication, in abstract, of the main discoveries already arrived at by his researches, and his memoir entitled, "Notice sur les fossiles du terrain crétacé du Jura Neuchâtelois" ("Mémoire Soc. Sc. nat. de Neuchâtel," Vol. I., p. 126, 1835) began his series of publications on "Echinides." He describes twelve species found in the cretaceous rocks of Neuchâtel, eight of which were entirely new. The paper is marked by great originality of classification, clearness of description, and exactness of drawing. It succeeded better than any other publication in showing that the cretaceous strata of Neuchâtel were a special formation, differing from the Green Sand and Gault of England, and, at the same time, younger than the Portland stone of the Jurassic system. At that time, the name *Neocomian*, to designate the Neuchâtel strata of the Lower Cretaceous had not yet been used by Thurmann, who offered it at a meeting of Jurassian geologists at Besançon, in September, 1835.

The "Prodrome d'une monographie des Radiaires ou Echinodermes," read also before the Neuchâtel Society on the 10th of January, 1834, was published in 1835, at page 168 of the first volume of the memoirs of that society; it is the starting-point of all the publications on the echinoderms, according to the principles of classification of Cuvier. Agassiz followed the method of his master; and in the first twelve pages he gives a most remarkable exposition of their zoölogical characters, and of his views on the classification and determination of the genera of that class of marine animals. Curiously enough, a year and a half later, in August, 1835, Charles Des Moulins, an able zoölogist of Bordeaux, without any knowledge of the works of Agassiz, published his "Études sur les Echinides," in three papers; Bordeaux, 1835-1837. The first two, published in August and December, 1835, contain no reference to Agassiz's researches; but the third paper, dated September, 1837, contests the *priority* of some of the genera created by Agassiz. As a whole, the two memoirs by Agassiz and

Des Moulins contain many similar results, and their coincidence of publication and of result, due to a mere accident, is very honourable to both. Des Moulins presented more facts and observations on the living and Tertiary echinoderms, while Agassiz brought forward more new forms and new genera from the secondary (Cretaceous and Jurassic) echinoderms, and also a better bibliographical knowledge of the subject.

The "Tableaux synonymiques des Echinides" of Des Moulins, is another "Prodrome," corrected and finished, according to Des Moulins himself, in April, 1837, more than a year after the publication of Agassiz's "Prodrome." Des Moulins says that when, on the point of finishing the manuscript of his "Tableaux synonymiques," he received from Agassiz a copy of his "Prodrome," and that he was thus enabled to interpolate all Agassiz's synonymy, including also the names of species described in works by several writers which Des Moulins did not possess and had never seen. The question of *priority* was settled by Des Moulins, in favour of Agassiz; Des Moulins's claims to priority being limited to two genera: *Collyrites* instead of *Disaster* of Agassiz, and *Echinocidaris* instead of *Arbacia* of Gray; a detail simply in Agassiz's classification of the great family of Echinoderms.

The winter of 1834–1835 and all the spring of 1835 were devoted to his great work on fossil fishes, the echinoderm studies being considered by Agassiz as a sort of relaxation and recreation. New numbers of the "Poissons fossiles" were issued, the text not corresponding with the atlas of plates, which at the time rendered rather difficult and confusing the task of those who wanted to follow him. His two friends of Berlin, Leopold von Buch and Alexander von Humboldt, complained of this, and von Buch went so far as to call his method of issuing text in fragments from different volumes *diabolical.* Humboldt, although calling Agassiz, in his letter of the 10th of May, 1835, "a great and profound naturalist," and speaking of his "admiration of your eminent works," adds: "I also complain a little, though in all humility; but I suppose it to be due to the difficulty of concluding any one family of (fossil fishes), when new materials are daily accumulating on your hands. Continue then as before. In my judgment, M. Agassiz never does wrong." To any one who knows how sarcastic and sharp Humboldt was, it is surprising to see him treating Agassiz so tenderly, using circumlocution in admonishing him, and placing the burden of sharpest criticism on his friend von Buch.

The isolation of Agassiz in a small town, beyond direct intercourse with other naturalists and savants in general, had already begun to tell. If he had been exposed to daily friction with his fellow-naturalists, he would have avoided many mistakes and false steps.

In July, 1835, Agassiz took his young wife on a visit to her parents at Carlsruhe, and left her there while he went a second time to England, where he remained until the end of October, working hard at descriptions of all the fossil fishes he had collected the previous year, and revising and directing the work of his two artists; for besides Dinkel, he now had a second draughtsman, M. Weber, another of his Munich friends. The expenses had grown so large that he began to think that he had "committed an imprudence in throwing myself into an enterprise so vast in proportion to my means as my 'Poissons fossiles.'" His publisher, Cotta of Stuttgart, had abandoned the undertaking as being too expensive and attended with too many *aleas*, and Agassiz bravely resolved to be his own publisher, — a very rash decision on his part, taking into account his complete lack of business capacity; but as he says: "Having begun it, I have no alternative; my only safety is in success. I have a firm conviction that I shall bring my work to a happy issue, though often in the evening I hardly know how the mill is to be turned to-morrow."

At the meeting of the British Association for the advancement of science in Dublin, which Agassiz attended, another appropriation of one hundred guineas, similar to the one voted the preceding year toward the facilitating of researches upon English fossil fishes, was granted him, which allowed him to pay his two artists. His presence in England and Ireland greatly helped the subscriptions to his work. English savants acted generously, and Agassiz's reputation grew rapidly among them. But, nevertheless, English enthusiasm never went so far as to offer him a single official position during his whole life.

In France the number of subscriptions was far below what it was in England, only fifteen copies being disposed of. Again, at this time, the loss of Cuvier was felt; for he alone would have had the power to get a subscription for fifty or sixty copies from the government, as he did for his "Poissons vivants," which would have placed Agassiz at ease. Properly engineered, Agassiz might have succeeded in getting the French government interested in his great work, but for some reason he withdrew from the undertaking, and did not even make an attempt in that direction during his stay in Paris.

An incident occurred at Dublin, during the meeting of the British Association, which was recorded in a letter from Adam Sedgwick to Lyell, dated Sept. 20, 1835. Sedgwick says: "Agassiz joined us at Dublin, and read a long paper to our section (the Geological Section). But what think you? Instead of teaching us what we wanted to know, and giving us of the overflowing of his abundant ichthyological wealth, he read a long, stupid, hypothetical dissertation on geology, drawn from the depths of his ignorance. And, among other marvels, he told us that each formation (*e.g.*, the lias and the chalk) was formed at one moment by a catastrophe, and that the fossils were by such catastrophes brought from some unknown region, and deposited where we find them. When he sat down, I brought him up again, by some specific questions about his ichthyological system, and then he both instructed and amused us. I hope we shall, before long, be able to get this moonshine out of his head, or at least prevent him from publishing it. His great work is going on admirably well. I think it is by far the most important work now on hand in the geological world" ("Life and Letters of Sedgwick," Vol. I., p. 447, Cambridge, 1890). Agassiz wisely withdrew his very objectionable paper. It was one of the weak points of his disposition to indulge in wild suppositions on subjects of which he knew very little, and to plunge into speculation absolutely out of his range of research.

It was on this occasion, at a festival at Florence Court, the seat of Lord Enniskillen, that Enniskillen, as it was related by his son, Lord Cole, to Lyell, was put "in great good humour," for long time after, by the perfect coolness with which Agassiz made "Murchison and some other guest *glorious*, and Sedgwick *comfortable*."[1] Such a jolly set of hammer-bearers Lord Enniskillen had never seen before, and Murchison acknowledged that he had found in Agassiz his master. At the hospitable table of Lord Enniskillen the old Munich student proved a match for the old trooper of the Peninsula War.

Not long after his return to Neuchâtel, a son, Alexander, so named in honour of Agassiz's best friend, Alexander Braun, was born on the 1st of December, 1835. For more than one reason it was a great event in the family, for from that moment Mrs. Agassiz, who showed herself at once an excellent and most careful mother, entirely abandoned pencil and books, and devoted all her time and strength to her son, and afterward to her two daughters, — one Ida, born Aug. 8, 1837; and the other, Pauline, born Feb. 8, 1841.

[1] "Life of Sir Charles Lyell," Vol. I., p. 457, and also "Life and Letters of Sedgwick," Vol. I., p. 445.

Indeed, with the scanty means at her disposal, Mrs. Agassiz had her hands full, and even more than full, as we shall see by and by; and it is not surprising that she could no longer manifest active interest in her husband's scientific work. It would have been beyond human power to continue her work of drawing fossil fishes and helping at manuscripts.

But we must not anticipate: let us return to Agassiz's various and constantly increasing work at Neuchâtel. Four letters written at this time to two naturalists, who were counted among his best and most trusted friends, Jules Pictet of Geneva and Célestin Nicolet of La Chaux-de-fonds, will give an intimate view of his scientific activity.

NEUCHÂTEL, 24 novembre, 1833.

MONSIEUR JULES PICTET,
à Genève.

Monsieur,—Je viens de recevoir votre lettre et je m'empresse d'y répondre, dans l'espoir d'obtenir le plus vite possible les objets que vous voulez bien offrir à notre Musée. J'espère que dès à présent, nous pourrons entrer en relations d'échanges suivies. M. Coulon et moi sommes dans ce moment occupés à ranger, et à déterminer les collections, pour en mettre les doubles à notre disposition, ce qui facilitera beaucoup nos échanges. . . . En échange nous pouvons vous offrir en général surtout des Poissons surtout plusieurs espèces d'eau douce nouvelles et inédites, des Mollusques en esprit de vin et des coquilles d'espèces vivantes, des coquilles fossiles surtout du Lias et des étages jurassiques inférieurs du Würtemberg, des Zoophytes en esprit de vin et des fossiles; des roches, surtout des séries complétes du Grés Bigaré, du Muschelkalk, du Keuper et des terrains jurassiques; les fossiles et les roches de la Craie des environs de Neuchâtel qui sont très nombreux. Nous avons aussi beaucoup de doubles des plantes d'Allemagne. Voilà donc assez de matériaux pour faire de nombreux échanges; veuillez seulement, s'il vous plait préciser davantage ce que vous désirez recevoir d'abord, et puisque vous voulez bien nous faire le premier envoi ne pas trop tarder à le faire. Si vous aviez, des *Diceras* et en général des fossiles de la Montagne des Fis (Savoie), vous nous obligeriez beaucoup de nous en envoyer, nous voudrions pouvoir les comparer avec notre Craie. Si vous avez des espèces de poissons du Brésil qui ne soient pas mentionnées dans mon ouvrage, elles seraient aussi bien venues pour notre Musée.

Je fais maintenant imprimer la 2ème livraison des "Poissons fossiles," qui contiendra la description des genres *Platysomus*, *Tetragonolepis*, *Dapedium*, *Lemionotus*, et *Lepidotus*, et une partie de l'Ostéologie générale des poissons. Il est fâcheux que les publications périodiques obligent les auteurs à morceler leurs sujets; mais enfin avec le temps on finit par les rendre complets.

Puisqu'enfin vous voulez bien m'offrir votre appui dans mes recherches sur les objets qui vous entourent de plus près, oserai-je vous prier de bien vouloir m'adresser par la Méssagerie, un jour qu'il fera froid; un exemplaire de votre Gravanche, et une ou deux de Féra, de différentes dimensions. Je vous offre en échange les Corégones du lac de Constance, de Bavière et de Neuchâtel. Je crois avoir vidé la question des Salmones d'Europe, ce n'est plus qu'à la synonymie que je dois donner encore quelques soins; aussi je cherche à receuillir tous les noms de province. Cuvier dans la 2ème édition du "Rêgne Animal" a admis beaucoup trop d'espèces.

Agréez, Monsieur, l'assurance de mon dévouement, et de ma considération distinguée.

LS. AGASSIZ.

NEUCHÂTEL, 22 avril, 1835.

MONSIEUR JULES PICTET,
à Genève.

Monsieur, — Depuis que j'ai eu le plaisir de correspondre avec vous pour notre premier échange, l'arrangement de nos poissons s'est très avancé et une grande partie de la collection est triée, et les doubles sont mis à part. Il nous est donc maintenant bien plus facile d'éffectuer les échanges que précédemment; cependant nos catalogues ne sont point encore faits. C'est pour cette raison, Monsieur, que tout en acceptant avec reconnaissance l'offre que vous me faites pour notre Musée, je vous prierais si cela pouvait vous convenir, de bien vouloir nous envoyer un exemplaire de toutes les espèces de poissons exotiques que vous possédez; en revanche, je vous adresserai tout ce que vous n'avez pas encore de mes poissons d'eau douce, et si cela ne suffit pas, j'ai encore quelques exemplaires de poissons des grandes rivières du Brésil. Enfin j'ai rapporté d'Angleterre une telle masse de fossiles, que je ne serais pas embarrassé de vous transmettre l'équivalent des poissons que vous me feriez parvenir. Si parmi vos espèces il s'en trouvait que nous eussions, ce dont je doute, je pourrais vous les renvoyer avec les nôtres.

S'il vous manque beaucoup de poissons de la Méditerranée, je pourrais vous en fournir beaucoup. Je désirerais également beaucoup connaître les poissons du lac de Lugano, du moins les espèces des genres critiques.

Je pense ne plus renvoyer d'un an, la publication de mes "Poissons d'eau douce"; la $5^{\text{èm}}$ livraison des (Poissons) fossiles paraîtra dans six semaines.

Agréez, Monsieur, l'assurance de ma considération distinguée, et de mon entier dévouement.

LS. AGASSIZ.

NEUCHÂTEL, le 4 Mars, 1834.

MONSIEUR CÉLESTIN NICOLET,
à La Chaux-de-fonds.

Monsieur, — J'ai reçu il y a 15 jours et lu le même soir à notre société la notice détaillée que vous nous avez adressée sur le calcaire lithographique des Montagnes (de Neuchâtel). Votre communication a excité l'intérêt qu'elle mérite et tous les membres de la société vous féliciteront de votre zèle si vous parvenez à découvrir quelque localité où l'on puisse lever des plaques assez grandes pour exécuter les travaux lithographiques.

Il se rattache une question géologique à vos recherches qui me parait importante sous le point de vue scientifique, c'est l'appréciation rigoureuse des rapports de position qui existe entre votre calcaire lithographique et celui de Sohlenhofen. En Bavière le calcaire est stratifié en couches horizontales et c'est surement de là que vient la beauté des pierres de Sohlenhofen; tandisque dans la chaine du Jura tous les calcaires ont été disloqués pas des soulèvements postérieurs à leur déposition et sont par conséquent très fendillés. Il serait bien intéressant d'avoir une collection un peu étendue des fossiles de votre calcaire afin de pouvoir les comparer avec le grand nombre de ceux que l'on trouve à Sohlenhofen; ce serait un moyen de plus de déterminer les relations géologiques de ces dépôts. Si vous avez occasion d'en receuillir, ne le négligez pas; ce serait un grand service que vous nous rendriez de nous en adresser le plus d'échantillons possibles.

Agréez, Monsieur, l'assurance de ma considération distinguée.

LS. AGASSIZ.

NEUCHÂTEL, le 19 Mars, 1835.

MONSIEUR CÉLESTIN NICOLET,
à La Chaux-de-fonds.

Monsieur, — La Société des Sciences Naturelles de Neuchâtel ayant décidé d'imprimer ses mémoires a nommé un comité pour en faire un choix et soigner l'impression. Ce comité désirant voir vos observations géologiques figurer dans son receuil m'a chargé de vous demander l'autorisation de faire imprimer votre notice sur la pierre lithographique des montagnes,[1] en vous priant d'y ajouter d'abord vos nouvelles observations sur les gisement de ces couches, sur leur âge géologique, et sur les fossiles qu'elles contiennent. Vous nous obligerez infiniment en répondant bientôt à notre appel. Déjà les premières feuilles de nos Mémoires sont imprimées.

J'ai beaucoup regretté de ne m'être pas trouvé à Neuchâtel lorsque vous vous y êtes réunis avec Messieurs Voltz, Thurmann et Thirria; mais j'espère avoir bientôt le plaisir de faire votre connoissance, M. Ladame m'ayant proposé il y a déjà quelques temps de faire une course avec lui à la Chaux-de-fonds.

Agréez, Monsieur, l'assurance de ma considération très distinguée.

LS. AGASSIZ.

[1] "Essai sur le calcaire lithographique des environs de La Chaux-de-fonds" ("Mémoires de la Soc. Sc. nat. de Neuchâtel," Vol. I., p. 66, 1835).

CHAPTER V.

1836–1837.

THE WOLLASTON MEDAL — FIRST PAPER OF DE CHARPENTIER ON THE GLACIAL THEORY — VENETZ'S OBSERVATIONS ON LARGE BOULDERS PERCHED ON THE SIDES OF THE ALPINE VALLEYS — DR. HERMANN LEBERT, THE FIRST DISCIPLE AND PUPIL OF DE CHARPENTIER AND VENETZ — EXTRACT FROM DE CHARPENTIER'S FIRST PAPER — AGASSIZ'S SUMMER VACATION AT BEX, NEAR THE HOUSE OF DE CHARPENTIER — CONVERSION OF AGASSIZ TO THE GLACIAL THEORY; HIS CREATION OF THE ICE-AGE — KARL SCHIMPER VISITS AGASSIZ AT BEX AND AT NEUCHÂTEL — DISCOURS DE NEUCHÂTEL JULY 24, 1837, ON THE ICE-AGE.

THE year 1836 was happily inaugurated by the reception of the Wollaston Medal, awarded to Agassiz by the Geological Society of London, at its annual meeting on February 19. The President, Charles Lyell, a life-long friend of Agassiz, in presenting the medal, said: —

On a former occasion we presented the proceeds of the Donation Fund[1] for one year to the same distinguished naturalist, to assist him in the publication of the early part of his great work, the importance of which was then only beginning to be known to the scientific world. It will ever be a subject of gratification to us to have learned that this small pecuniary aid was not without its influence in accelerating the publication of his "Researches on

[1] The sum of thirty guineas, or £31, 10*s.* sterling.

Fossil Fish," arriving as it did opportunely at a moment when the funds which could be appropriated for the undertaking were nearly exhausted. Mr. Agassiz acknowledged at the time his obligation to us for a mark of sympathy and regard which he received so unexpectedly from a foreign country, and which cheered and animated him to fresh exertions. The Council, in now awarding the Medal to him, are desirous that he should possess a lasting testimony of their esteem and of the high sense which they entertain of the merit of his scientific labours.

It was a well-deserved reward, received when quite a young man,—in his thirtieth year only,—which did honour to the Geological Society of London as well as to the recipient. Never since has the Wollaston Medal been bestowed on so young a naturalist; his is a unique case, and as such is recorded on the List of Awards of the Wollaston Medal.

In 1834, at the meeting of the Helvetic Society of Naturalists, at Lucerne, Jean de Charpentier, Director of the Salt Works at Bex, Canton de Vaud, had read a short paper entitled, "Notice sur la cause probable du transport des blocs erratiques de la Suisse." Seldom, if ever, has such a small memoir so deeply excited the scientific world. It was received at first with incredulity and even scorn and mockery, Agassiz being among its opponents. Its publication, however, a year later, and again eighteen months later, in the "Annales des Mines" of Paris, Vol. VIII., p. 219, and in the "Bibliothèque universelle" of Geneva, Vol. IV., p. 1, with a German translation by Julius Froebel and Oswald Heer, in "Mittheil. aus dem gebiete der theoret. erdkunde," p. 482, Zürich, attracted much attention, and the smile of incredulity with which it was received when read at Lucerne soon changed into a desire to know more about it.

A mountaineer, Perraudin, of the Bagnes valley, at the foot of the St. Bernard, in Valais, told de Charpentier, as far back as 1815, that the large boulders perched on the sides of the Alpine valleys were carried and left there by glaciers. De Charpentier thought the hypothesis so extraordinary and extravagant that it was not worth examining or even considering. Fourteen years later, in 1829, at the meeting of the Swiss naturalists at the Grand St. Bernard's Hospital, his good and most esteemed friend, the engineer of the "Ponts et Chaussées" of the Valais Canton, M. Venetz, not only supported the view advanced by Perraudin, but told the Society that his observations led him to believe that the whole Valais has been formerly covered by an immense glacier, and that it even extended outside of the canton, covering all the "Canton de Vaud" as far as the Jura Mountains, carrying all the boulders and erratic materials, which are now scattered all over the large Swiss valley. In 1821 the extremely modest Venetz had read before the Swiss naturalists a paper entitled, "Mémoire sur les variations de la température des Alpes de la Suisse." In some way the memoir was left entirely unnoticed, and the manuscript put aside. De Charpentier, as soon as he was convinced of the correctness of the Venetz theory, hunted up the manuscript, which was buried under the dust of the archives of the Helvetic Society of Naturalists, and had it finally printed and published, in 1833, — twelve years after it was written, — in " Erstern Bandes zweyte abtheilung," of the "Denkschiften der allgemeine Schweizerischen Gesellschaft für die gesammten Naturwissenchaften."

As this initial memoir on the glacial epoch is extremely rare, I will quote the conclusions and one paragraph:—

> Monsieur Perraudin, conseiller de la commune de Bagnes, habile chasseur de chamois, et amateur de ces sortes d'observations [on old moraines], nous a assuré que les glaciers de Séveren, de Loui, et de la Chaux-de-Sarayer, tous dans la vallée de Bagnes, ont des moraines fort reconnaissables, qui sont environ à une lieue de la glace actuelle. . . .
>
> Nous sommes donc en quelques manières autorisés à croire:—
>
> 1) Que les moraines qui se trouvent à une distance considérable des glaciers, datent d'une époque *qui se perd dans les nuits des temps*. . . .
>
> 6) Que les glaciers parviendront difficilement à la hauteur gigantesque, dont nous trouvons tant de vestiges. . . .

Civil Engineer Venetz was not educated as a scientific man, and he did not understand the scientific method of marshalling and classifying facts and observations. But he found in his friend de Charpentier the best possible man to systematize and construct a new science. If it was Venetz who developed and sustained the hypothesis of the chamois hunter, Perraudin, and awaked de Charpentier's interest in the question, it was de Charpentier, who by his scientific method of observation, his clear and logical reasoning, accumulated and classified on truly scientific bases all the material proofs, such as the *moraines*, the *roches moutonnées polies et striées*, the *cailloux striés*, the *boue glaciaire*, etc., and to de Charpentier is due the glacial doctrine and the glacial theory.

As early as 1833, de Charpentier had gathered round

him at Bex, pupils and believers in the new science, including among the first ones O. Heer, afterward so celebrated for his researches in fossil botany and fossil entomology, E. Thomas, the botanist, and the learned Dr. H. Lebert. The latter, a brilliant German political refugee from Breslau, an enthusiastic friend and great admirer of de Charpentier, who justly compared the splendid and characteristic profile of de Charpentier to that of Keppler and of Galileo, and pronounced his head as typical of a savant, came to Bex in August, 1833, and was there convinced of the soundness of the views of Venetz and de Charpentier. For him the beautiful demonstrations of de Charpentier were conclusive, and left no doubt; so much so that in 1834, on the occasion of his receiving his degree of Doctor of Medicine at Zürich, and before de Charpentier had read his paper at Lucerne, he gave a public lecture on the glacial theory.

The just and honest Heer, in his "Le Monde primitif de la Suisse," has nobly upheld the claims of de Charpentier, saying: "C'est Jean de Charpentier qui le premier donna une base scientifique à cette hypothèse par une série de recherches consciencieuses et par une rigoureuse combinaison des faits connus." And he further says: "Jean de Charpentier est le fondateur de la théorie des glaciers."

The short paper of de Charpentier contains some of the fundamental principles on which the glacial theory is based, and is so important that some extracts will be acceptable to all those who like to follow the history of a great discovery from its infancy.

Extracts from NOTICE SUR LA CAUSE PROBABLE DU TRANSPORT DES BLOCS ERRATIQUES DE LA SUISSE; *par M. J. de Charpentier, Directeur des mines du canton de Vaud.* (*Extrait du Tome VIII des "Annales des Mines," pp.* 20. *Paris,* 1835.)

M. Venetz, en étudiant les glaciers, a été conduit à s'occuper des blocs erratiques transportés par la vallée du Rhône, et l'examen qu'il a fait de ces blocs, et des diverses circonstances qui les accompagnent, l'a convaincu que leur transport n'a pas pu s'effectuer par le moyen de l'eau, quelque énormes qu'on suppose son volume et sa vitesse, et quelque puissante que soit son action. . . .

Les dépôts des blocs erratiques présentent constamment un mélange informe de fragmens de toutes les dimensions, depuis celle d'un grain de sable jusqu'à celle de plusieurs milliers de pieds cubes. On trouve sur le Jura des blocs aussi volumineux que dans les vallées des Alpes. Il n'existe donc point de triage selon les volumes et les poids relatifs des blocs, ce qui nécessairement aurait dû avoir lieu s'ils avaient été entraînés et amenés par l'eau; car, dans ce cas, les plus gros blocs devraient se trouver les plus voisins du lieu d'où la débâcle et le courant les auraient enlevés, et ces fragmens devraient diminuer de volume à mesure qu'ils en sont plus éloignés, de manière que les blocs qu'on trouve sur les pentes du Jura devraient être en général sensiblement plus petits que ceux qu'on rencontre au pied et dans les vallées des Alpes. Mais, nous le répétons, un pareil arrangement ne s'observe nulle part. . . .

Quoique la plupart des blocs erratiques présentent une forme arrondie évidemment par frottement, on en trouve néanmoins qui sont non-seulement aplatis, mais qui sont restés presque intacts, ayant à peine leurs angles et leurs arêtes écornées ou émoussées. Si leur déplacement avait eu lieu par un courant, on ne saurait pas concevoir comment ils auraient pu être roulés jusques au pied du Jura et poussés sur son faîte, sans porter des marques violentes de frottement.

Les dépôts de ces roches transportées présentent ordinairement une forme alongée, semblable à celle d'une digue ou d'un rempart, ou bien ils forment quelquefois des monticules coniques, isolés ou disposés en file. Ils ne se rencontrent jamais en forme de nappe ou de plateau. Ces digues sont placées horizontalement au pied et sur la pente des montagnes, ordinairement les unes derrière les autres, et espacées à des distances inégales : elles sont parallèles entre elles et à la direction de la vallée. Quelquefois deux ou plusieurs de ces digues se trouvent tellement rapprochées les unes des autres, qu'elles se confondent en une seule, terminée par une ou plusieurs arêtes. La plus grande élévation à laquelle on les trouve sur la pente des montagnes qui bordent la vallée du Rhône, est d'environ 1.100 à 1.200 pieds au-dessus de ce fleuve, dans les environs de Bex, et de 2.400 pieds dans ceux de Sion. Le sol sur lequel ils reposent n'est jamais formé d'atterrissemens ou d'éboulemens, mais c'est toujours du roc en place.

La disposition et la configuration extérieure de ces dépôts sont inexplicables par la théorie d'un transport par le moyen d'un courant d'eau ; car l'eau les aurait déposés en forme de nappes, surtout dans les plaines des vallées et dans celles qui se trouvent au pied des Alpes ; cette théorie n'explique pas non plus comment ces blocs auraient pu franchir, sans les combler, les lacs qui se trouvent à l'extrémité inférieure de la plupart de nos grandes vallées, ni la singulière position de ces énormes blocs qu'on trouve isolés dans la plaine ou sur la pente des montagnes, plantés verticalement sur le sol, et quelquefois brisés ou fendus du bas en haut dans toute leur longueur, ce qui semble indiquer qu'ils sont tombés à peu près verticalement sur la place même où ils se trouvent encore, et qu'ils se sont fendus ou brisés par leur chute.

On remarque en outre que les blocs sortis d'une vallée latérale ne se mêlent point ou très imparfaitement avec ceux de la grande vallée, ou avec ceux qui sont sortis d'une vallée opposée. Ainsi les pierres feldspathiques ou talqueuses de la vallée d'Hérens, formant des dépôts considérables près de Sion, ne se mêlent point avec les blocs calcaires qui proviennent des vallées de la Sionne et de la Lierne, qui toutes les deux prennent naissance auprès de Rawyl, et se terminent à la grande vallée du Rhône, à peu près vis-à-vis de la vallée d'Hérens. Les digues ou remparts qu'imitent les dépôts de blocs de chacune de ces vallées sont parfaitement séparés et distincts.

Feu M. Escher de la Linth avait déjà remarqué ce même fait par rapport aux grandes vallées de la Suisse, c'est-à-dire que les blocs erratiques de la vallée du Rhône ne se mêlaient point avec ceux qui étaient sortis de la vallée de l'Aar ; que ces derniers restaient séparés et distincts des dépôts de blocs venus de la vallée de la Reuss, etc. En admettant un courant d'eau ou une débâcle qui ait eu lieu instantanément et à la fois dans ces diverses vallées, on ne comprend pas pourquoi et comment les pierres entraînées ne se mêlaient pas dans les endroits où ces courans venaient se toucher et se joindre, et surtout là où ils frappaient contre le Jura, ce qui aurait dû produire une sorte de remou ou de refoulement, qui loin d'empêcher le mélange des matériaux que ces courans amenaient avec eux, l'aurait au contraire singulièrement favorisé.

Un autre phénomène qu'on observe dans les vallées de toutes les chaînes de montagnes qui ont fourni des blocs erratiques, ce sont les surfaces lisses que présentent les rochers qui n'ont pas été dégradés par la décomposition ou par des éboulemens. Ces surfaces sont évidemment le résultat d'un frottement, et comme on sait que les eaux courantes qui charrient du sable et des pierres, usent et polissent les rochers avec lesquels elles viennent en contact, on a cru que les surfaces lisses et usées des rochers de nos grandes vallées étaient dues à la débâcle ou au grand courant qu'on supposait avoir transporté les blocs erratiques, qui, en quelque sorte, auraient fait office de l'émeril. Pour donner plus de probabilité à cette explication, on alléguait le fait incontestable que ces surfaces polies ne se rencontrent pas audessus du niveau que les blocs transportés ont atteint de chaque côté de la vallée, et qu'au-dessus de ce niveau les rochers n'offrent que des surfaces raboteuses, de véritables cassures.

La supposition d'une débâcle ou d'un courant n'explique pas d'une manière satisfaisante ce phénomène ; car comment concevoir qu'une si immense quantité de blocs de toutes les dimensions, mise en mouvement par une énorme masse d'eau, ait pu unir et rendre lisses des surfaces verticales et d'une grande étendue? Loin de les polir, elle n'aurait fait que les écorner et les ébrécher. Comment des blocs entraînés par l'eau auraient-ils pu frotter et user des surfaces qui surplombent, qui forment ces sortes de voûtes que nos montagnards désignent par le nom de *barmes* ou de *balmes?* Comment expliquer, par cette supposition, la formation de surfaces polies, derrière des rochers qui font saillie, et qui, par ce fait même, auraient dû les préserver du courant, et les protéger contre le choc et le frottement des corps solides charriés par l'eau?

Mais laissons de côté ces difficultés et admettons pour un moment que ces surfaces lisses avaient été produites par un courant d'eau; dans ce cas elles devraient être plus marquées vers l'extrémité inférieure des vallées que dans leur partie supérieure ou vers leur naissance, et elles devraient être absolument nulles sur les côtés des Alpes. Eh bien, c'est précisément le contraire; ces surfaces lisses et polies se recontrent depuis le pied jusqu'au faîte des Alpes, et plus on s'élève, mieux on les trouve prononcées; elles sont extrêmement distinctes sur le Saint-Bernard, le Simplon, le Saint-Gothard, le Grimsel, la Gemmi, le Sanetsch, le col d'Enzeindaz, etc. . . .

Je pourrais citer encore d'autres faits plus ou moins contraires à la théorie d'un courant d'eau, si ceux que je viens d'indiquer ne me paraissaient pas suffire pour faire soupconner que l'agent qui a transporté les blocs erratiques a été tout autre qu'une débâcle ou une masse d'eau en mouvement.

M. Venetz croit que des *glaciers* ont été cet agent, et que ces dépôts de blocs erratiques ne sont autre chose que des *moraines*.

Je sens fort bien tout ce qu'une pareille hypothèse offre au premier abord d'invraisemblable, de choquant, d'extravagant même. En effet, comment admettre, comment se persuader que jadis toutes nos grandes vallées fussent occupées dans toute leur longueur par de vastes glaciers, qui, à leur débouché dans la plaine au pied des Alpes, se seraient étendus en forme de nappes ou d'énormes éventails pour couvrir presque toute la contrée jusqu'au Jura, et remonter cette chaîne en nombre d'endroits jusques à son faîte, et le dépasser même? Comment concilier une semblable hypothèse avec la masse de faits qui prouvent que jadis la température de nos climats a été bien plus élevée qu'elle ne l'est maintenant? . . .

J'avoue que toutes ces objections et beaucoup d'autres se présentèrent à moi lorsque M. Venetz, il y a environ cinq ans, me fit part de son opinion. Je restai dans le doute, jusqu'à ce que les faits que j'avais mis tant de soin à rechercher et à examiner pour combattre cette hypothèse, m'eussent conduit à un résultat tout opposé à celui auquel je m'étais attendu. . . .

Les glaciers et leurs moraines se placant devant l'entrée de quelque petite vallée latérale, y forment une sorte de barre, qui, empêchant l'écoulement des eaux, change le vallon en lac, dans lequel les torrens amènent des pierres, des sables et des limons, et les déposent par lits. Il n'est donc pas surprenant de rencontrer quelquefois auprès des dépôts des blocs erratiques de petits amas de matériaux évidemment stratifiés et déposés par l'eau.

Quoique la plupart des blocs charriés par les glaciers soient arrondis, ou aient au moins leurs angles et leurs arêtes émoussés ou écornes par le frottement qu'ils éprouvent les uns contre les autres, néanmoins on trouve quelquefois sur le dos des glaciers de gros blocs isolés qui arrivent sans frottement, et par conséquent bien conservés jusqu'au pied du glacier. Ce fait explique la manière dont quelques-uns des blocs erratiques ont pu être transportés à de fort grandes distances sans avoir éprouvé de frottement, et sans que leurs angles et leurs arêtes aient été sensiblement endommagés.

La forme des moraines est celle d'une digue ou d'un rempart, terminé par une ou plusieurs arêtes. Dans certains cas elle est conique, ou bien elle présente une foule de monticules coniques. Lorsqu'un glacier, comme il arrive le plus souvent, a plusieurs moraines, elles sont toujours parallèles entre elles, et placées à des distances inégales. La configuration intérieure et extérieure des moraines, et leur disposition mutuelle sont donc exactement les mêmes que celles des dépôts des blocs erratiques.

Les glaciers ne produisent jamais, comme les torrens et les rivières, de dépôts en forme de lits ou de nappes, parce qu'ils creusent toujours le terrain jusqu'au roc vif, poussant devant eux toutes les terres, graviers et blocs qu'ils rencontrent sur leur passage, phénomène connu de tous ceux qui ont observé des glaciers dans le temps où ils sont en progression, et qui s'explique très bien par la manière dont les glaciers augmentent et avancent. Puisque les glaciers en s'avançant déblaient le terrain jusqu'au roc vif, nous pouvons facilement concevoir pourquoi nos lacs n'ont pas été comblés par la quantité immense de blocs, de gravier et de sable qui ont dû les traverser, ou plus exactement, qui ont dû passer par dessus, et qui, s'ils avaient été amenés par de l'eau, n'auraient pas manqué de les remplir. . . .

Depuis les travaux de M. de Saussure, tout le monde sait que deux glaciers, lorsqu'ils viennent à s'atteindre et à se joindre sous un angle aigu, ne mêlent et ne confondent point leurs moraines. Ce fait explique parfaitement pourquoi les blocs erratiques d'une de nos grandes vallées ne sont point mêlés avec ceux de la vallée voisine, phénomène duquel on ne saurait se rendre compte par la supposition que le transport de ces blocs eût été opéré par le moyen de l'eau. . . .

On sait que les glaciers frottent, usent et polissent les rochers avec lesquels ils sont en contact. Cherchant à s'étendre, ils suivent toutes les sinuosités, et se pressent et se moulent en quelque sorte dans tous les creux et toutes les excavations qu'ils peuvent atteindre, et en polissent les surfaces, même celles qui surplombent, ce qu'un courant d'eau charriant des pierres ne pourrait effectuer.

Comme les glaciers prennent naissance sur le faîte des Alpes, leur action destructive doit avoir duré beaucoup plus long-temps dans les régions supérieures que dans les basses vallées et à leur pied. Il n'est donc pas étonnant de rencontrer dans les hautes vallées et sur les cols des Alpes des marques de frottement beaucoup plus considérables et mieux prononcées que vers leur pied, ce qui devrait être précisément l'inverse si ce frottement avait été opéré par un courant ou une débâcle. Enfin l'observateur qui part du faîte du Jura dans la direction même où les blocs erratiques y sont arrivés, en suivant constamment leur trace, se trouve conduit jusqu'au fond des hautes vallées des Alpes, et jusqu'aux glaciers qui les dominent, où il voit enfin ces dépôts devenir de véritables moraines. . . .

Je termine cette notice en exprimant le vœu qu'elle puisse attirer l'attention des naturalistes sur le travail que prépare M. Venetz; qu'elle puisse les engager à étudier derechef le grand phénomène des blocs erratiques.

Agassiz resolved to pass his summer vacation of 1836 in a healthy locality among the Alps. At that time resorts were few, and there were none at all in the centre of the Alps. At about the time of his marriage, in 1833, de Charpentier had invited him to visit him at his beautiful home "aux Dévens," near Bex. De Charpentier, the classmate, at the Freiberg School of Mines, of Alexander von Humboldt and Leopold von Buch, the author of the best geological description of the Pyrenees then existing, had a European reputation which brought to his house savants from every country; in addition, he enjoyed the reputation of a charming and most hospitable companion, and was the possessor of rich collections of natural history. De Charpentier had married, in 1828, a young German lady of noble family, Miss von Gablenz of Dresden; and as Mrs. Agassiz was not particularly fond of Swiss ladies, Agassiz thought that an acquaintance with Mrs. de Charpentier, a German lady of culture and refinement, might be agreeable to his wife.

It is a mistake to think that Agassiz was attracted to Bex by a desire to study the glacial question. He was adverse to the hypothesis, and did not believe in the great extension of glaciers and their transportation of boulders, but, on the contrary, was a partisan of Lyell's theory of transport by icebergs and ice-cakes. His main object was to pass an agreeable vacation with his wife and child, at the foot of the Dent du Midi, and near a family of savants as social and friendly as were de Charpentier and wife. In all this he was not disappointed; but from being an adversary of the glacial theory, he returned to Neuchâtel an enthusiastic convert to the views and observations of Venetz and de Charpentier. Agassiz found lodgings in the neighbourhood of "des Dévens," at "la Sallaz," a suburb of the small town of Bex, and daily visited de Charpentier. The site, just north of Bex, on rising ground, among fine orchards and vineyards, is truly magnificent; with luxuriant vegetation, and in full view of the opening of the great valley of the Valais, and at the foot of the Dent du Midi. Mrs. Agassiz, with her little boy Alexander, was delighted with the place, and with Mr. and Mrs. de Charpentier, as well as their only child, a charming girl of seven years, time passed quickly, and Agassiz found in more intimate acquaintance with de Charpentier, another charmer of men, not like himself in many points, but very similar in some. For instance, de Charpentier was a delightful talker, very hospitable, and, like Agassiz, enjoyed hearing the "chime at midnight." The evenings passed like dreams, in endless conversations on scientific subjects. For the greater comfort of the guests collected round his table, — for besides Agassiz there were Dr. Lebert, Em. Thomas, Venetz, Albert Mousson, Escher von der Linth, and Lardy, — de Charpentier ordered the best wine of his cellar, and although moderation prevailed, the conversation was often enlivened, and hour after hour passed so quickly that the company frequently did not separate until a late hour; sometimes not before daybreak. It was a fruitful and genial time for all those who were fortunate enough to be present. Agassiz was soon converted into a glacialist by the arguments, and more especially by the evidences shown him by de Charpentier and Venetz, all round Bex, and in several excursions to the Valais. With his power of quick perception, his unmatched memory, his perspicacity and acuteness, his way of classifying, judging, and marshalling facts, Agassiz promptly learned the whole mass of irresistible arguments collected patiently during seven years by de Charpentier and Venetz, and with his insatiable appetite and that faculty of assimilation which he possessed in such a wonderful degree, he digested the whole doctrine of the glaciers in a few weeks.

Agassiz saw also that de Charpentier was a true "scientific epicurean" in the best and most elevated sense of the word, as he had been characterized by Dr. Lebert, not only without ambition for fame, but even indifferent as to the diffusion of his discoveries among scientific men. Lebert calls de Charpentier "une Belle au bois dormant"; and it was for Agassiz to play the rôle of the prince in awaking him, and obliging him to publish his researches; which he finally did in October, 1840, under the title of "Essai sur les Glaciers."

Agassiz, with his extraordinary imagination, saw that the phenomenon of the extension of old glaciers had not been confined to the Rhone valley, but must have been general, and formed a special period in the history of the earth, during which cold prevailed all over the world. In a word, Agassiz's sojourn at Bex, under the teaching of de Charpentier, had taught him, with his far-reaching thoughts, to add an entirely unexpected, and, at that time generally very unacceptable, stage to the various periods which the earth had passed through; namely, the Ice-age.

On his return to Neuchâtel, Agassiz began to examine attentively, with the new tool he had obtained at Bex, all the vicinity of Neuchâtel and Bienne, finding everywhere the most unmistakable proofs of glacial action, and of the extension of the glacier of the Rhone to the Chaumont, with its "Pierre à Bot," and far away north towards Soleure.

During his stay at Bex, Agassiz, as a good friend, wished to share the great pleasure afforded to him by his stay near de Charpentier, and he kindly invited Karl Schimper to visit him. As Agassiz said in 1842, in his defence against the attacks of Schimper, "Through the highly interesting works of Venetz and de Charpentier upon glaciers, my attention was called to these phenomena. In the autumn of 1836 I went to Bex, where I spent several months, and under the guidance of M. de Charpentier gradually learned to understand these remarkable phenomena." These plain words cannot leave any doubt as to the fact that Agassiz became converted to the glacial theory by the teaching of de Charpentier. Schimper, who did not leave Bex with Agassiz at the beginning of November, but accepted the hospitality tendered to him by de Charpentier, was not with Agassiz when he made his observations on the polished and scratched rocks and boulders round Neuchâtel. After lingering several weeks at de Charpentier's hospitable and generous house, Schimper rejoined Agassiz at Neuchâtel as his guest, as he had been at Bex and formerly at Munich. Of course, being constantly together, Agassiz and Schimper carried on a continual exchange of views on the Ice-age. During the winter of 1836-37, Agassiz gave a public lecture at Neuchâtel on the subject. He was continually haunted by his thoughts on old glaciers; and when the Helvetic Society of Natural Sciences, of which he had been elected president, met at Neuchâtel on the 24th of July, 1837, he wrote during the night previous his famous "Discours d'ouverture." In it Agassiz most frankly acknowledges that his explanation of the glacial epoch "est le résultat de la combinaison de mes idées et de celles de M. Schimper." All these explanations are necessary, in order to show exactly how Schimper became involved in the question, and how unjust are the accusations of plagiarism launched against Agassiz by Schimper himself and by Dr. Otto Vogel, in the "Allgemeine Zeitung" of Augsburg. Agassiz's good heart and constant readiness to give impulse to new ideas were interpreted in a manner not exactly creditable.

But before we come to the delivery of his "Discours," let us see how friendly he was to Schimper. As soon as Schimper became a guest in Agassiz's apartment at Neuchâtel, Agassiz introduced him to everybody and made the most of him. At the meetings of the Neuchâtel Society of Natural Sciences, Schimper communicated in February, March, and April, 1837, his observations on the morphology of plants, showing the laws of development of leaves round the axes; and also his new ideas on the development of the animal kingdom before the appearance of man. During the five years which had passed since their last meeting at Carlsruhe in 1832, Schimper had undergone changes which were not to his advantage. He had failed to draw a line sharply separating his student life from his life as privatdocent and instructor at the Munich University. His appetite, on the contrary, had greatly developed, and was almost beyond his control. However, the society of de Charpentier at Bex and of Agassiz at Bex and Neuchâtel was beneficial to him; and he never was so brilliant and attractive.

On the 15th of February, 1837, which was the anniversary of his birth, Schimper was in particularly excellent spirits. That evening he first made two verbal communications before the Natural History Society on botanic morphology, promising to write them for the "Bulletin," — a promise which, by the way, was never carried into effect, like all Schimper's promises, — and then he distributed to Agassiz and all the friends there a small piece of poetry, half-scientific, half-humorous, in which, for the first time, the word *Eiszeit* (glacial epoch), so celebrated since, was printed. Schimper had the honour to be the god-father of a great geologic period, for it was certainly he who first coined and used the word. Agassiz always acknowledged his *priority*; and on the 25th of July, before the geological section of the Helvetic Society, he read a letter from Schimper, addressed to him under the title of "Ueber die Eiszeit," in which the word *Eiszeit* is written in italics, and so printed on p. 38 of the "Actes de la Société Helvetique," Neuchâtel, 1837.

But poor Schimper soon fell again into bad habits after leaving Agassiz, and the brilliant spirit, the rare genius, — for a man of genius Schimper certainly was, — became more and more obscured, until he disappeared entirely, without leaving even a good manuscript account of his great discovery on the morphology of plants.

The "Discours de Neuchâtel" is the starting-point of all that has been written on the "Ice-age." Quoted often, it is, however, very little known, because it never was printed separately and also because the number of copies of the small volume of the "Actes de la Société Helvétique réunie à Neuchâtel" was extremely limited. As it occupies such an important place in the history of the progress of geology, and also in the life of Agassiz, I think it is proper to reproduce it *in extenso* and in French, as it was delivered.

Discours prononcé à l'ouverture des séances de la Société Helvétique des sciences naturelles, à Neuchâtel, le 24 *Juillet,* 1837, *par L. Agassiz, Président.*

Messieurs, très chers amis et confédérés :

Depuis longtemps les membres de la section neuchâteloise de notre société désiraient avec impatience voir arriver le moment où ils pourraient inviter leurs confrères de toute la Suisse à se réunir chez eux. Des circonstances indépendantes de leur volonté, et particulièrement la construction du nouvel édifice dans lequel nous sommes réunis et qui devait recevoir tout ce que la ville possède de collections scientifiques, les ont forcés à décliner l'honneur d'acceuillir à Neuchâtel la Société Helvétique des sciences naturelles, jusqu'à ce qu'ils pussent le faire convenablement et mettre sous ses yeux au moins une partie des collections. Encore aujourd'hui, malgré toute l'activité qu'y a mise l'infatigable Directeur de notre Musée, il n'y a qu'une faible partie des collections qui soient rangées ; c'est même à la hâte qu'elles ont été déposées dans le local qui doit les recevoir et que les ouvriers n'ont pas encore quitté. Nous réclamons donc toute votre indulgence pour ce que vous verrez. Mais du moins, comptez sur le plaisir que nous avons à vous recevoir ici, et soyez persuadés que nous attachons un grand prix à vous voir chez nous. C'est du fond du cœur que je vous dis à tous: Soyez les bien-venus.

A pareil jour tout nous invite à rechercher quel est le lien qui unit les sciences dont s'occupe notre Société. Je ne crois pas me tromper en affirmant qu'une grande pensée domine tous les travaux qui tendent aujourd'hui à en étendre les limites. C'est l'idée d'un développement progressif dans tout ce qui existe, d'une métamorphose à travers différens états dépendant les uns des autres, l'idée d'une création intelligible, dont notre tâche est de saisir la liaison dans tous ses phénomènes.[1] Ainsi voyez l'Astronomie, qui s'occupe maintenant de la formation des corps célestes; la Chimie, qui étudie les différens modes d'action des corps les uns sur les autres; la Physique, qui veut approfondir la nature des forces dont elle connaît l'action; l'Histoire Naturelle, qui poursuit les phases de la vie de chaque être; la Géologie enfin, qui se hasarde à embrasser l'histoire de la terre, à en déchiffrer même les pages les plus anciennes, et à la représenter comme un grand tout, dont les révolutions ont toujours tendu vers le même but.

De tous ces progrès, sans doute, il sortira un jour quelque chose de grand, de vraiment *humain*, qui fera rentrer l'étude des sciences naturelles bien plus directement dans le domaine de la vie habituelle de l'homme, que les avantages mêmes fournis à l'industrie et aux arts par les résultats obtenus dans les sciences, quelques immenses qu'aient été ces derniers.

Notre Société n'est point restée étrangère à ce grand mouvement; les noms de ses membres figurent honorablement à côté des coryphées de la science qui ont daigné s'associer à nos travaux. La réunion d'aujourd'hui, mieux qu'aucune autre peut-être, prouverait que mon assertion n'est point exagérée. Vous le savez, Messieurs, c'est notre petite société qui a servi de modèle à ces vastes associations dont l'Allemagne, l'Angleterre, et la France se glorifient à

[1] If Agassiz had replaced the words "développement progressif" and "métamorphose" by *evolution*, what a splendid Darwinian paragraph he would have given there, in 1837, twenty-two years before the publication of the "Origin of Species." — J. M.

tant de titres ; et si les travaux qu'elle a enterpris ont paru moins brillans, à côté de ceux de sociétés plus vastes, elle n'en a pas moins donné l'élan, à plus d'une reprise.

Tout récemment encore, deux de nos collègues ont soulevé par leurs recherches des discussions d'une haute portée, et dont les suites auront du retentissement. La nature de la localité où nous sommes réunis m'engage à vous entretenir de nouveau d'un sujet qui, je crois, trouve sa solution dans l'examen des pentes de notre Jura. Je veux parler des glaciers, des moraines, et des blocs erratiques.

Tout le monde, en Suisse, connaît les glaciers et sait que leurs bords sont entourés de digues de blocs arrondis qu'on appelle des *moraines*, et qui sont continuellement poussées en avant ou abandonnées par les glaciers à mesure qu'ils avancent ou qu'ils se retirent. Les habitans du Jura surtout sont familiers avec un autre phénomène qui est très frappant dans nos montagnes, je veux parler des *blocs erratiques* ou de ces masses de granit et d'autres roches primitives qui sont éparses principalement sur les pentes de notre Jura. Ce que tout le monde ne sait cependant pas, c'est qu'il existe encore d'autres moraines que celles qui cernent de nos jours les glaciers. Ce sont MM. Venetz et de Charpentier, qui les ont fait connaître les premiers. On les observe principalement dans les vallées inférieures des Alpes. Mais il est un côté de cette question qui doit être contesté, c'est la liaison que l'on a cherché à établir entre les blocs erratiques et les glaciers que cernaient les grandes moraines dont on retrouve encore des traces sur les rives septentrionales du lac de Genève. C'est de ce dernier point que j'ai l'intention de vous entretenir en particulier.

Les *faits* observés par MM. Venetz et de Charpentier sont cependant définitivement acquis à la science ; aussi importe-t-il d'en proclamer hautement l'exactitude ; car de là dépend naturellement la validité de toutes les conséquences que l'on peut en tirer.[1]

A des distances plus ou moins considérables des glaciers actuels, on remarque en effet à différentes hauteurs des moraines parfaitement semblables à celles qui cernent encore les glaciers. Elles sont

[1] It is impossible to say more clearly, or with more force, that Messrs. Venetz and de Charpentier founded the glacial doctrine. J. M.

également concentriques et forment des digues qui suivent les inégalités des flancs des vallées. On en voit partout plusieurs étages, dont les plus élevés se trouvent à quelques cents pieds au-dessus du fond des vallées supérieures des Alpes où il n'y a plus de glaciers. Mais en descendant dans les vallées *inférieures*, on en trouve successivement à douze ou quinze cents pieds et même à plus de dix-huit cents pieds de hauteur; il y en a encore d'assez distinctes à deux mille pieds au-dessus du lit du Rhône, dans les environs de St. Maurice en Valais. On peut les poursuivre jusque sur les rives du lac de Genève. Il en existe encore de très-élevées au-dessus de Vevey et dans les environs de Lausanne, qui correspondent à celles de la rive méridionale du lac.

Si on ne les a généralement pas remarquées, c'est qu'elles sont beaucoup au-dessus des routes fréquentées, et que celles des parties inférieures des vallées ont généralement été disloquées par les torrens.

Il est toujours facile de distinguer ces anciennes moraines des digues formées par le débordement des eaux et des talus plus ou moins étendus, résultant des avalanches. Les digues sont très-irrégulières et s'étendent à de petites distances, en s'aplanissant; les talus sont en forme de cônes très-aplatis, débouchant des vallées et se perdant dans la plaine; tandis que les moraines sont des digues triangulaires continues et parallèles le long des deux flancs des vallées, formées de blocs arrondis évidemment triturés, pour ainsi dire en place, les uns contre les autres, comme cela a lieu sur le bord des glaciers actuels, qui s'étendent dans de longues vallées étroites. Les blocs des avalanches, au contraire, sont anguleux; ceux des digues, charriés par les eaux, peuvent être arrondis, il est vrai, lorsqu'ils proviennent de moraines disloquées, mais alors ils s'étendent en *nappes* irrégulières, et lorsqu'ils proviennent d'avalanches récentes, ils sont également anguleux, à moins qu'ils ne rencontrent dans leur trajet d'anciennes moraines qu'ils entraînent et avec lesquelles ils se confondent.

Pour se convaincre de l'exactitude de ces faits, il suffit de parcourir la vallée de Chamouni, en suivant les moraines les plus rapprochées des glaciers, ou de s'élever perpendiculairement sur les flancs de la vallée du Rhone entre St. Maurice et Martigny, sur la rive gauche du Rhone, au-dessus de la Pissevache près du hameau appelée Chaux-Fleurie (Tsau-fria), ou vis-à-vis en montant au village de Morcles depuis les bains de Lavey. Les décombres des dernières débâcles de la Dent du Midi, les grandes avalanches dont on voit partout des traces et les nombreuses digues formées par le Rhône, feront d'ailleurs apprécier justement la différence qu'il y a entre ces divers accidens produits par des causes si différentes.

Les vallées latérales présentent les mêmes phénomènes, comme on peut le voir en remontant le cours de l'Avençon, jusqu'au glacier de Paneyrossaz.

En parcourant ces vallées, je n'ai pas été moins frappé de l'apparence polie que présentent les rochers sur lesquels les glaciers se sont mus; apparence que l'on remarque également dans toutes les vallées dont les flancs sont couronnés d'anciennes moraines, à quelque distance des glaciers actuels qu'elles se trouvent. C'est ainsi que les flancs de la vallée du Rhône sont entièrement polis jusque sur les bords du lac de Genève à plus d'une journée des glaciers, partout où la roche est assez dure pour avoir résisté aux influences atmosphériques.

L'explication que M. de Charpentier a donnée de ces faits, évidemment produits par de grandes masses de glaces, qui remplissaient jadis le fond de toutes les vallées alpines, ne me semble cependant pas embrasser toute la question, et le Jura présente une série de phénomènes qui la mènent plus loin.

Pour mettre plus de liaison dans ce que j'ai à vous dire là-dessus, je vous entretiendrai d'abord des surfaces polies que l'on remarque sur toute la pente méridionale du Jura et que nos montagnards appellent des *laves*, comme nous l'a appris M. Léopold de Buch, celui de tous les géologues qui le premier a le mieux étudié le Jura Neuchâtelois et à qui sont dus les plus grands travaux sur le sujet qui nous occupe.

La pente méridionale du Jura, qui est en face des Alpes, présente de ces *laves* jusque sur ses plus hautes sommités, depuis les bords du lac de Bienne jusqu'au delà d'Orbe; limites dans lesquelles j'ai constaté leur existence.[1] Ce sont des surfaces polies, complètement

[1] Elle s'étendent cependant bien au-delà, comme nous l'apprend une lettre de M. Schimper, reçue le 25 Juillet et insérée à la page 38 de ces Actes.

indépendantes de la stratification des couches et de la direction de la chaîne du Jura; elles s'étendent sur toute la surface du sol, suivant ses ondulations, passant également par dessus le terrain néocomien et le terrain jurassique, pénétrant dans les dépressions qui forment de petites vallées, en s'élevant sur les crêtes les plus isolées et présentant un poli aussi uni que la surface d'un miroir, partout où la roche a été mise récemment à découvert, c'est-à-dire, débarrassée de la terre, du gravier et du sable qui la recouvrent généralement. Ces surfaces sont tantôt planes, tantôt ondulées, souvent même traversées de sillons plus ou moins profonds et sinueux, ou de bosses longitudinales très-arrondies, mais qui ne sont jamais dirigés dans le sens de la pente de la montagne; au contraire, comme les gibbosités, ces sillons sont obliques et longitudinaux; direction qui exclut tout idée d'un courant d'eau comme cause de ces érosions. Un fait très-curieux, que l'on ne saurait non plus concilier avec l'action de l'eau, c'est que ces polis sont uniformes, alors même que la roche se compose de fragmens de différente dureté, et les coquilles qu'elle contient sont tranchées comme dans des plaques de marbre polies artificiellement. On remarque, en outre, sur les surfaces très-bien conservées de fines lignes semblables aux traits que pourrait produire une pointe de diamant sur du verre, et qui suivent en général la direction des sillons obliques. Les localités les plus intéressantes où l'on peut les observer dans les environs de Neuchâtel, sont le Mail, du côté du lac, à la surface du terrain néocomien, et le Plan, à l'endroit où l'ancienne route joint la nouvelle. Les plus remarquables sont cependant à quelque distance de la ville, par exemple, au-dessus du Landeron, à la surface du portlandien sur la lisière des vignes et de la forêt, dans les environs de St. Aubin et au-dessus de Concise. Dans quelques localités on remarque de larges excavations et même des espèces de puits qui ne peuvent avoir été produits que par des cascades tombant entre les fentes de la glace. Pour quiconque a examiné dans les Alpes le fond des anciens glaciers, il est évident que c'est la glace qui a produit ces polis, comme ceux de la vallée du Rhône dont il a déjà été question. Il est digne de remarque que ces polis ne se retrouvent nulle part dans le fond des petites vallées longitudinales formées par les abruptes des différentes ceintures des couches dont se composent nos chaînes, ni sur l'escarpement même de ceux de ces abruptes qui sont tournés vers la montagne, tandisque j'en ai remarqué sur plusieurs abruptes tournés vers les Alpes, par exemple, le long de la route neuve entre St. Aubin et le château de Vauxmarcus. Il importe également de signaler les différences qui existent entre ces *laves* et d'autres surfaces polies avec lesquelles on ne saurait cependant les confondre, mais qui peuvent leur ressembler dans quelques circonstances. Je veux parler des surfaces polies produites par les failles ou par le glissement des couches les unes sur les autres. Les premières pénétrant verticalement ou obliquement à travers plusieurs couches, ne sont à découvert que là où l'un des côtés de la roche en rupture s'est enfoncé; elles ne sont jamais à découvert sur de grandes surfaces comme les laves; les secondes présentent quelquefois des surfaces assez étendues, lorsque les couches supérieures au glissement ont été enlevées; mais alors les rainures ou les sillons produits par le glissement, sont dans le sens de la pente, ce qui ne se voit nulle part à la surface des laves. Les surfaces polies par l'action des eaux ont également un caractère particulier, soit qu'elles aient été produites par des eaux courantes ou par des masses d'eau plus considérables contenues dans un bassin. Dans le premier cas, ce sont des sillons sinueux descendant toujours, tandisque les sillons et les gibbosités des laves montent et descendent suivant les accidens de la roche polie. Dans le second cas, les eaux mues sur les rivages par les vents, et poussées au-delà de leur niveau habituel, rentrant toujours en équilibre, forment des sillons inégaux plus ou moins profonds, qui suivent généralement la ligne de plus grande pente, à moins que des accidens locaux ne leur donnent une direction particulière. Il en est de même lors de la hausse et de la baisse du lac au printemps et en automne. On peut étudier toutes ces différences dans les environs de la ville, en comparant les surfaces polies *du Mail* avec les érosions produites par le lac dans le prolongement des mêmes couches, ou avec les sinuosités qui ont été produites par le Seyon dans ses gorges. D'ailleurs les surfaces polies par l'action de l'eau ne sont jamais aussi lisses que les laves ou que les surfaces polies par les glaciers. Que l'eau charrie du sable et du limon ou non, les effets sont les mêmes, seulement ils sont plus lents dans ce dernier cas. Je n'ai pas encore eu occasion d'étudier particulièrement les effets des grandes masses d'eau charriant des glaces; je ne pense cependant pas qu'elles produisent des effets différens de ceux de l'eau liquide. Ce qu'il y a de certain, c'est que dans les lits de nos rivières et sur les bords de nos lacs ces effets se confondent; et puis il est évident que la glace flottante ne saurait avoir d'action sur le fond de l'eau qui la porte. Il n'y a donc que les grandes masses de glaces se mouvant immédiatement sur des masses solides qui puissent produire des effets semblables au poli que l'on remarque sur les bords des glaciers en retraite. Ce dernier phénomène est du reste parfaitement semblable à celui que présentent les laves du Jura.

Par cette ressemblance seule on pourrait déjà être porté à penser que des causes semblables ont produit des effets aussi semblables entr'eux. Mais il est d'autres considérations qui nous permettent de lier plus directement ces deux phénomènes, et qui forceront, même ceux qui voudraient y voir des agens différens, à les envisager sous un seul et même point de vue.

Nous avons vu des moraines jusques sur les bords du lac de Genève, sur les deux rives à la même hauteur; nous avons par-là la certitude qu'il fut un temps où le lac de Genève était gelé jusqu'au fond, et où cette glace s'élevait à une hauteur très-considérable au-dessus de son niveau actuel.

Mais nous savons également que toutes les moraines qui restent en place sont celles que les glaciers laissent sur leurs bords en se retirant. Depuis l'époque donc que je viens de signaler et où les glaciers débouchaient encore dans les vallées inférieures de la Suisse, ils sont allés en diminuant et en se retirant dans des vallées de plus en plus élevées.

Ici une question se présente tout naturellement. Ceux de ces glaciers qui ont eu la plus grande extension, sont-ils descendus du sommet des Alpes? ou bien y aurait-il eu un moment où les glaces se seraient formées naturellement au-delà des limites que nous venons de leur reconnaître, s'étendant peut-être une fois jusqu'au Jura et même au-delà?

Le niveau des moraines des bords du lac Léman, qui sont à 2500 pieds au-dessus de la mer, et la nature des surfaces polies du Jura semblent l'indiquer; il suffit même de marquer sur une carte de nivellement les hauteurs des moraines débouchant dans les différentes parties de la chaîne des Alpes, pour se convaincre que les glaces ont une fois recouvert toute la plaine de la Suisse et atteint la pente du Jura.[1]

En effet, la différence de niveau entre l'élévation des moraines des bords du lac de Genève aux environs de Vevey et sur la côte de Savoie, et celle des surfaces polies que l'on observe au-dessus des rivages du lac de Neuchâtel jusque sur le sommet de Chaumont, est telle que la nappe de glace qui remplissait l'espace compris dans ces limites, a pu avoir une certaine inclinaison, puisque le niveau du lac de Neuchâtel n'est que de 1344 pieds au-dessus de la mer, celui de la zône de Pierre-à-Bot, le long de laquelle on trouve le plus grand nombre de blocs, de 2150 pieds ; le sommet même de Chaumont n'a que 3619 pieds.

Cela étant, nous sommes non-seulement en droit d'attribuer à l'action des glaces toutes ces surfaces polies de la pente du Jura, mais encore de les envisager comme un indice assuré de l'étendue plus considérable qu'ont eue les glaces à une époque plus reculée, tant dans le Jura que dans les Alpes.

M. de Charpentier pense que ces glaces étaint des glaciers qui se sont formés sur le sommet des Alpes et qui sont descendus dans la plaine pour s'élever jusqu'à la hauteur où on en trouve des indices, poussant devant eux les blocs qui sont sur le Jura. Mais un fait bien frappant s'oppose à cette explication ; c'est que les blocs du Jura sont généralement moins arrondis et même plus grands que ceux

[1] M. Rod. Blanchet, qui s'est aussi occupé de cette question, a fait dès lors la remarque que le sommet du Pélerin (montagne qui domine Vevey en face de l'ouverture du Valais, élevée de 3301 pieds de France au-dessus de la mer), composé de poudingue à gros grain, est poli sur sa pente, dans un endroit où il n'y a pas d'eau capable de former un petit ruisseau, ni de sentier, ni aucune des causes polissantes que l'on pourrait mettre en avant.

C'est donc à 3300 pieds au moins que l'on peut porter le niveau des glaces qui remplissaient le bassin du lac de Genève, dont la surface n'est maintenant qu'à 1145 pieds. Sur le sommet du Pélerin c'est le *fond* de la grace dont le niveau était de 3300 pieds au-dessus de la mer ; mais rien ne nous indique quelle était son épaisseur dans ce point.

que l'on trouve dans les moraines du bord des glaciers actuels.[1] Si nos blocs avaient été roulés ainsi au bord d'un glacier depuis les Alpes jusqu'au Jura, ils seraient généralement plus ronds, et plus petits, et il y aurait d'immenses moraines adossées au Jura, ce qui n'est pas.[2]

L'opinion généralement reçue attribue le transport de ces blocs à d'immenses courans d'eau ou à des glaces flottantes.

Les plus grandes difficultés que présente cette manière de voir, pour n'en indiquer que quelques-unes, sont d'abord d'expliquer l'origine de ces courans et de la vitesse qu'on doit leur attribuer pour qu'ils aient pu transporter des masses aussi énormes, si toutefois l'on admet qu'ils ont été charriés *après* le soulèvement des Alpes, comme tout semble l'indiquer. Car dans ce cas, ces courans auraient dù partir des *crêtes* qui séparent les vallées, puisque le phénomène des blocs se présente dans toutes les vallées alpines et sur les deux versans de la chaîne, c'est-à-dire que pour suffire aux exigences des faits, ils auraient dû jaillir de toutes ces crêtes[3] avec assez d'impétuosité pour ne plus laisser tomber les blocs au-dessous du niveau où ils se trouvent dans le Jura et dans les vallées alpines où ils n'y a plus de glaciers, puis qu'on nie même encore l'existence

[1] Ces faits ne s'accordent point du tout avec ceux que M. Élie de Beaumont a décrits pour la vallée de la Durance.

[2] Je ne me suis point attaché à décrire la distribution des blocs erratiques sur les pentes du Jura, parce qu'elle est assez connue depuis la publication des recherches de MM. Léop. de Buch, Escher de la Linth, de Luc, sur ce sujet. Je ferai seulement remarquer que leur accumulation sur différens points ne s'accorde pas avec les théories que l'on a avancées pour expliquer leur transport. Ainsi les plus grandes accumulations que j'en connaisse se trouvent à peu de distance l'une de l'autre près du sommet du mont Auber, et dans le fond de Noiraigue, à des niveaux très-différens, et qui ne sont point sur une ligne ascendante dont le sommet serait à Chasseron. Au contraire, c'est en général sur le bord des différens gradins du Jura qu'on en voit le plus, et en particulier sur la lisière que forme tout le long du Jura Neuchâtelois, la dépression des couches supérieures du portlandien, entre le château de la Neuveville, Fontaine-André, Pierre-à-Bot, Troirod, Châtillon, Fresens, Mutruz, etc.

[3] Les systèmes de barrage et de débâcles que l'on pourrait imaginer, n'expliqueraient jamais des faits communs à tant de vallées à la fois.

des grandes moraines, pour attribuer aussi la déposition de ces blocs aux mêmes courans. Mais comment des cours d'eau ayant à peine quelques lieues de long (je parle ici des vallées latérales débouchant dans les vallées principales) auraient-ils maintenu de grands blocs à plus de mille pieds de hauteur? D'ailleurs le fait que les blocs des différentes vallées ne sont pas les mêmes et qu'ils se répandent *en éventail* à une certaine distance des Alpes, exclut cette idée d'une extrême vitesse qu'on a voulu accorder aux courans, uniquement pour expliquer le transport des blocs, sans penser qu'ils auraient dû produire en même temps d'autres effets dont on ne retrouve aucune trace. Ce fait exclut à plus forte raison l'idée d'-un grand courant diluvien passant sur toute la Suisse, quelque direction qu'on veuille lui assigner. Si c'est *avant* le soulèvement des Alpes qu'on suppose que le phénomène a eu lieu, je demande comment il se fait que les lignes que ces blocs forment dans les Alpes n'ont pas été disloquées par le soulèvement? car dans ce cas les digues continues et parallèles de blocs que l'on voit *sur les deux flancs* de toutes les vallées alpines et qui en suivent tous les accidens, quelles que soient leur direction et leurs sinuosités, restent inexplicables, l'eau suivant un cours rectiligne dans les différentes anfractuosités du lit qu'elle parcourt, tandis que la glace seule agit avec la même énergie sur tous les points des bassins qu'elle remplit.

Les objections que l'on peut faire contre la théorie des courans sont toutes applicables jusqu'à un certain point à la théorie de M. Lyell, d'un charriage par des glaces flottantes. On peut bien faire arriver par des radeaux de glaces des blocs anguleux jusque sur le Jura; mais les autres particularités de ce grand phénomène ne s'expliquent pas plus par là, qu'à l'aide des courans, dût-on même admettre avec M. Élie de Beaumont que leur eau provenait de la fonte des glaciers.

Une autre objection d'un très-grand poids faite à cette théorie par M. Schimper, c'est l'état actuel des lacs et de la grande vallée suisses. Si les blocs ont été charriés par des courans depuis les Alpes au Jura, ces courans ont naturellement passé par dessus les lacs et les vallées longitudinales et transversales qui se trouvent entre deux. Comment se fait-il alors que ces lacs et ces vallées n'ont point été comblés? et comment expliquer les escarpemens anguleux de leurs bords?

Quelque violens, quelque rapides, quelque profonds que l'on suppose ces courans, eussent-ils même, contre toutes les lois de la physique, porté des blocs de granit d'environ 50,000 pieds cubes, comme celui de Pierre-à-Bot, ils ont dû se ralentir une fois, et alors les dernières traînées auraient encore dû combler quelques-unes de ces inégalités. Cependant on voit peu de blocs entre les Alpes et le Jura.

Si dans une autre hypothèse on les fait marcher lentement sur des masses de limon et de décombres assez épaisses pour les porter, comment se fait-il que ces masses du moins n'ont pas comblé toutes les inégalités de la Suisse? Les blocs seuls se seraient-ils peut-être déposés après être arrivés sur le Jura, et les masses qui avaient pu les apporter jusques là se seraient-elles alors écoulées pour les laisser en place?

D'autres considérations s'opposent encore à l'admission de tous ces courans.

Les blocs erratiques du Jura reposent partout sur des surfaces polies, à moins qu'ils n'aient été poussés au-delà des crêtes de nos montagnes, et qu'ils ne soient tombés dans le fond des vallées longitudinales, comme on le voit dans toute la vallée du Creux-du-Vent. Mais ce n'est pas *immédiatement* sur les surfaces polies qu'ils sont gisant. Partout où les cailloux roulés qui accompagnent les grands blocs n'ont pas été remaniés par des influences postérieures, on remarque que les petits blocs, des galets de différente grandeur, forment une couche de quelques pouces et quelquefois même de plusieurs pieds, *sur* laquelle les grands blocs anguleux reposent. Ces cailloux sont de plus très-arrondis, même polis et entassés de manière à ce que les plus gros soient dessus les plus petits qui passent souvent à un fin sable au fond, immédiatement sur les surfaces polies. Cet ordre de superposition, qui est constant, s'oppose à toute idée d'un charriage par des courans; car dans ce dernier cas, l'ordre de superposition des cailloux arrondis serait inverse. La présence d'un fin sable à la surface des roches polies, prouve en outre qu'aucune cause puissante n'a agi, ou qu'aucune catastrophe importante n a atteint la surface du Jura, depuis l'époque du transport de ces roches alpines, ou en d'autres termes, que les surfaces polies lors du transport des blocs n'ont pas été disloquées depuis. Mais comme ces surfaces forment en grande partie la rive septentrionale des lacs de Neuchâtel et de Bienne, elles prouvent, pour eux du moins, que les lacs suisses existaient déjà; et la continuité des moraines sur les deux rives du lac de Genève, prouve que ce bassin aussi est antérieur au transport des blocs, puisqu'il a précédé la formation des moraines, comme on le verra bientôt.

En considérant la liaison intime des différens faits qui viennent d'être décrits, il est évident que toute explication qui ne rendra pas compte en même temps du poli de la surface du sol, de la superposition et de la forme arrondie des cailloux et du sable qui reposent immédiatement au-dessus des surfaces lisses, et de la forme anguleuse des grands blocs superficiels, est une explication inadmissible pour les blocs erratiques du Jura; et c'est le cas de toutes les hypothèses sur le transport des blocs que je connais.

Voici quelle est l'explication de tous ces phénomènes que je crois maintenant la plus plausible. Elle est le résultat de la combinaison de mes idées et de celles de M. Schimper sur ce sujet. En effleurant plusieurs questions générales qui s'y rattachent, pour chercher à l'établir, je n'ai point l'intention de les traiter à fond maintenant. Je veux simplement faire voir par là que le sujet qui nous occupe touche aux plus grandes questions de la géologie.

L'étude des fossiles porte depuis quelque temps des fruits bien inattendus, surtout depuis qu'elle a pris un caractère physiologique, c'est-à-dire depuis que l'on a entrevu qu'il existe un développement progressif dans l'ensemble des êtres organisés qui ont vécu sur la terre, et que l'on a reconnu des époques de renouvellement dans leur ensemble. Ceux qui ont compris ce progrès ne doivent pas craindre maintenant d'en poursuivre les conséquences jusques dans leurs dernières limites, et l'idée d'une diminution uniforme et constante de la température de la terre, telle qu'elle est admise, est tellement contraire à toute notion physiologique, qu'il faut la repousser hautement pour faire place à celle d'une diminution de température accidentée en rapport avec le développement des êtres organisés qui ont paru et disparu les uns à la suite des autres à des époques déterminées, se maintenant à une moyenne particulière pendant une époque donnée, et diminuant à des époques fixes.

Comme le développement de la vie individuelle est toujours accompagné de celui de la chaleur, que sa durée établit un certain équilibre plus ou moins durable, et que sa fin produit un froid glacial, je ne crois donc pas sortir des conséquences que les faits permettent de déduire, en admettant que sur la terre les choses se sont passées de la même manière : que la terre, en se formant, a acquis une certaine température très-élevée, qui est allée en diminuant à travers les différentes formations géologiques ; que pendant la durée de chacune d'elles, la température n'a pas été plus variable que celle de notre globe depuis qu'il est habité par les êtres qui s'y trouvent, mais que c'est aux époques de disparition, de ses habitans qu'a eu lieu la chute de la température, et que cette chute a été au-dessous de la température qui signale l'époque suivante et qui s'est relevée avec le développement des êtres apparaissant nouvellement.

Si cette manière de voir est vraie, et la facilité avec laquelle elle explique tant de phénomènes inexplicables jusqu'ici me fait penser qu'elle l'est ; si cette manière de voir, dis-je, est vraie, il faut qu'il y ait eu à l'époque qui a précédé le soulèvement des Alpes et l'apparition des êtres vivant maintenant, une chute de la température bien au-dessous de ce qu'elle est de nos jours. Et c'est à cette chute de la température qu'il faut attribuer la formation des immenses masses de glace qui ont dû recouvrir la terre partout où l'on trouve des blocs erratiques avec des roches polies comme les nôtres. C'est sans doute aussi ce grand froid qui a enseveli les Mammouths de Sibérie dans les glaces, congelé tous nos lacs, et entassé de la glace jusqu'au niveau des faîtes de notre Jura qui existaient avant le soulèvement des Alpes.

Cette accumulation de glace au-dessus de tous les bassins hydrographiques de la Suisse se concoit aisément quand on pense que les lacs une fois gelés jusqu'au niveau de leurs débouchés, les eaux courantes ne s'écoulant plus, et celles du ciel accrues par les vapeurs des régions méridionales qui, dans des circonstances pareilles devaient se précipiter abondamment vers le Nord, en ont rapidement augmenté l'étendue et rehaussé le niveau jusqu'à la hauteur qui a été constatée par les faits déjà énoncés. L'hiver de la Sibérie s'était établi pour un temps sur une terre jadis couverte d'une riche végétation et peuplée de grands mammifères, dont les semblables habitent de nos jours les chaudes régions de l'Inde et de l'Afrique. La mort avait enveloppé toute la nature dans un linceul, et le froid arrivé à son plus haut degré, donnait à cette masse de glace, au maximum de tension, la plus grande dureté qu'elle puisse acquérir. Lorsqu'on a été fréquemment témoin de la congélation d'un lac, on sait combien la glace est résistante dans cet état, et à quelle immense distance des corps durs jetés à sa surface peuvent y glisser par suite même d'une faible impulsion.

L'apparition des Alpes, résultat du plus grand des cataclysmes qui ont modifié le relief de notre terre, a donc trouvé sa surface couverte de glace, au moins depuis le pôle Nord, jusque vers les bords de la Méditerrannée et de la mer Caspienne. Ce soulèvement, en rehaussant, brisant, fendillant de mille manières les roches dont se compose le massif qui forme maintenant les Alpes, a également soulevé les glaces qui le recouvraient; et les débris détachés de tant de fractures et de ruptures profondes se répandant naturellement sur la surface inclinée de la masse de glace appuyée contre elles, ont glissé sur sa pente jusqu'aux points où ils se sont arrêtés, sans s'arrondir, puis qu'ils n'éprouvaient aucun frottement les uns contre les autres et qu'en se heurtant ils se repoussaient facilement sur une pente aussi lisse; ou bien, après s'être arrêtés, ils ont été portés jusques sur les bords ou dans les fentes de cette grande nappe de glace, par l'action particulière et les mouvemens propres à l'eau congelée, lorsqu'elle subit les effets des changemens de température, de la même manière que les blocs de rocher tombés sur des glaciers sont poussés sur leurs bords, par suite des mouvemens continuels qu'éprouve leur glace en se ramollissant et en se congelant alternativement aux différentes heures de la journée et dans les différentes saisons. Ces effets devraient être décrits en détail, mais comme ils sont en partie connus, je ne m'y arrête pas.[1] Je me borne à dire que la puissance d'action qui en résulte pour la glace est immense; car ces masses se mouvant continuellement sur elles-mêmes et sur le sol, broient et arrondissent tout ce qui y est mobile, et polissent les surfaces solides sur lesquelles elles reposent, en même temps

[1] M. Schimper a fait un beau travail sur les effets de la glace, auquel je renverrais mes lecteurs s'il était publié.

que leurs bords poussent devant eux tout ce qu'ils rencontrent, avec une force irrésistible. C'est à ces mouvemens qu'il faut attribuer la superposition étrange des cailloux roulés et du sable, qui reposent immédiatement sur les surfaces polies ; et c'est sans doute à la pression de ce sable sur les surfaces polies que sont dues les fines lignes qui s'y trouvent gravées, et qui n'existeraient pas si le sable avait été mu par un courant d'eau : car ni nos torrens, ni l'eau fortement agitée de nos lacs, ne produisent rien de semblables sur les mêmes roches. Quant à la direction longitudinale de ces fines lignes et des sillons que l'on remarque sur les surfaces polies, je ferai observer qu'elle a dû résulter de la plus grande facilité que devait avoir la glace à se dilater dans le sens de la grande vallée suisse, plutôt que transversalement, encaissée comme elle l'était entre le Jura et les Alpes ; ce phénomène n'ayant dû commencer qu'avec le retrait de la glace, à une époque où les Alpes étaient déjà debout. Je ne mets pas en doute, que la plupart des phénomènes attribués à de grands courans diluviens, et en particulier ceux que M. Seefström a fait connaître récemment, n'aient été produits par les glaces.

Lors du soulèvement des Alpes, la surface de la terre s'est réchauffée de nouveau, et la chaleur dégagée de toutes parts a dès-lors commencé à faire fondre ces masses de glaces, qui se sont successivement retirées jusques dans leurs limites actuelles. Des crevasses se sont formées d'abord dans les endroits où la glace était le plus mince, c'est-à-dire sur le sommet des montagnes et des collines qui en étaient recouvertes, puis le long des points les plus saillans de la plaine ; des vallées d'érosion ont alors été creusées au fond de ces crevasses, dans des localités où aucun courant d'eau ne pourrait couler sans être encaissé dans des parois congelées ; et quand la glace eut complètement disparu, les grands blocs anguleux qui couvraient sa surface, ou qui étaient tombés dans ses fentes, se sont trouvés sur un lit de petits cailloux arrondis, sous lesquels on trouve encore ordinairement un sable plus fin. En baissant de niveau, la glace a nécessairement dû occuper plus longtemps les dépressions du sol, les petites vallées longitudinales formées par les différentes ceintures des couches du Jura et le fond des lacs ; et c'est sans doute à ce fait qu'il faut attribuer la position bizarre de tant de blocs perchés à peine en équilibre sur les pointes les plus éminentes des rochers, et leur absence constante dans les enfoncemens, où on n'en trouve du moins que là où de nouvelles dilatations momentanées de la glace en retraite a pu les y précipiter.

Aussi longtemps que le niveau des glaces dans le Jura ne fut pas tombé au-dessous de la ligne de Pierre-à-Bot, les blocs qui étaient encore répandus sur toute sa surface, purent continuer à être poussés contre le Jura; mais bientôt après les glaces devenant fort minces sur toute la plaine suisse, durent en disparaître promptement et ne plus laisser que des taches dans les vallées profondes et dans les bassins des lacs, c'est-à-dire, qu'elles se trouvèrent bientôt resserrées dans les vallées inférieures des Alpes.

En réfléchissant à ce qui a dû se passer pendant cette retraite des glaces, on est naturellement porté à penser que le transport des cailloux roulés de la vallée du Rhin et la déposition du Löss en ont été un des premiers effets, d'autant plus que ces cailloux sont les mêmes que ceux qui se trouvent avec nos blocs, et que le Löss est évidemment le résultat du détritus de la molasse. De fréquentes débâcles ont pu alors seulement charrier aussi des blocs sur des radeaux de glaces à de très-grandes distances, ou même en entraîner quelques-uns plus loin dans leur courant.

La fonte et la macération des glaces et leur congélation réitérée dans les jours froids, ont produit beaucoup d'autres effets géologiques difficiles à expliquer par d'autres causes. Sans rappeler les vallées d'érosion, je pourrais citer ces sillons profonds qui ne sont pas des fissures et qui sont dominés par de grandes étendues de plaines; ou bien ces petits lacs qui se forment quelquefois sur le bord des glaciers, et qui remanient les roches menues accumulées sur leurs bords, de manière à leur donner une apparence stratifiée; ou bien les phénomènes analogues que l'on observe sur les limites des différentes stations où les grandes nappes de glace ont dû s'arrêter successivement dans leurs retraites, ou bien la dispersion des os des mammifères de l'époque diluvienne, sans qu'ils soient ni roulés, ni brisés, etc., ou encore une foule d'autres particularités qui ne peuvent avoir d'intérêt que lorsqu'on a embrassé l'ensemble de la question.

Dès ce moment la surface de la terre a dû être soumise de nouveau aux influences du cours régulier des saisons; ce fut alors le premier printemps des animaux et des plantes qui vivent de nos jours; les glaces s'étaient retirées jusqu'aux pieds des Alpes, du sommet desquelles il commençait à leur venir de nouveaux renforts. Mais bientôt elles subirent leurs dernières retraites en oscillant toujours, gagnant tantôt en étendue et poussant des blocs devant elles, tantôt se retirant dans des limites de plus en plus étroites. À chaque pied de terrain qu'elles abandonnaient, elles laissaient derrière elles, comme les glaciers actuels en retraite, quelques-unes de ces longues digues de blocs qui dominent encore les vallées alpines. Bientôt les lacs se dégelèrent aussi, les eaux prirent leur cours actuel, les vallées des Alpes furent balayées, et il ne resta plus de glace des frimats passés que sur les sommets de nos blanches montagnes.

Ce serait donc une grave erreur de confondre les glaciers qui descendent du sommet des Alpes, avec les phénomènes de l'époque des grandes glaces qui ont précédé leur existence.

Le phénomène de la dispersion des blocs erratiques ne doit donc plus être envisagé que comme un des accidens qui ont accompagné les vastes changemens occasionnés par la chute de la température de notre globe avant le commencement de notre époque.

Admettre une époque d'un froid assez intense pour recouvrir toute la terre à de très-grandes distances des pôles d'une masse de glace aussi considérable que celle dont je viens de parler, est une supposition qui paraît en contradiction directe avec les faits si connus qui démontrent un refroidissement considérable de la terre depuis les temps les plus reculés. Rien cependant ne nous a prouvé jusqu'ici que ce refroidissement ait été continuel, et qu'il se soit opéré sans oscillations; au contraire, quiconque a l'habitude d'étudier la nature sous un point de vue physiologique, sera bien plus disposé à admettre que la température de la terre s'est maintenue sans oscillations considérables à un certain degré, pendant toute la durée d'une époque géologique, comme cela a lieu pendant notre époque, puis qu'elle a diminuée subitement et considérablement à la fin de chaque époque, avec la disparition des êtres organisés qui la caractérisent, pour se relever avec l'apparition d'une nouvelle création au commencement de l'époque suivante, bien qu'à un degré inférieur à la température moyenne de l'époque précédente; en sorte que la diminution de la température du globe pourrait être exprimée par la ligne suivante :

Ainsi l'époque de grand froid qui a précédé la création actuelle, n'a été qu'une oscillation passagère de la température du globe, plus considérable que les refroidissemens séculaires auxquels les vallées de nos Alpes sont sujettes. Elle a accompagné la disparition des animaux de l'époque diluvienne des géologues, comme les Mammouths de Sibérie l'attestent encore, et précédé le soulèvement des Alpes et l'apparition des êtres vivans de nos jours, comme le prouvent les moraines et la présence des poissons dans nos lacs. Il y a donc scission complète entre la création actuelle et celles qui l'ont précédée ; et si les espèces vivantes ressemblent quelquefois à s'y méprendre à celles qui sont enfouies dans les entrailles de la terre, on ne saurait cependant affirmer qu'elles en descendent directement par voie de progéniture, ou, ce qui est la même chose, que ce sont des espèces identiques.

Partant de ce qui précède, on parviendra aussi un jour à déterminer quelle est l'époque géologique à laquelle le soleil a commencé à exercer une influence assez considérable sur la surface de la terre, pour y produire les différences qui existent entre ses zônes, sans que ces effets fussent neutralisés par l'action de la chaleur intérieure, à laquelle la terre a dû pour un temps une température très-uniforme sur toute sa surface.

Cette manière de voir, je le crains, ne sera pas partagée par un grande nombre de nos géologues qui ont sur ce sujet des opinions arrêtées ; mais il en sera de cette question comme de toutes celles qui viennent heurter des idées reçues depuis longtemps. Quelque opposition qu'on puisse lui faire, toujours est-il que les nombreux faits nouveaux relatifs au transport des blocs que je viens de signaler, et que l'on peut étudier si facilement dans la vallée du Rhône et aux environs de Neuchâtel, ont amené la question sur un autre terrain que celui sur lequel elle a été débattue jusqu'à présent.

Quand M. de Buch affirma pour la première fois, en face de l'école formidable de Werner, que le granit est d'origine plutonique, et que les montagnes se sont élevées, que dirent les Neptunistes? — Il fut d'abord seul à soutenir sa thèse, et ce n'est qu'en la défendant avec la conviction du génie qu'il l'a fait prévaloir. Heureusement que dans les questions scientifiques, les majorités numériques n'ont jamais décidé de prime abord aucune question.

La forme que j'ai donnée aux observations que je viens de présenter, éloignera, je l'espère, d'ici toute discussion sur ce sujet, à moins qu'on ne réclame qu'il en soit autrement. Cependant, comme je ne saurais espérer d'avoir convaincu de la vérité de ces vues ceux qui viennent de les entendre pour la première fois, je pense que la section de Géologie sera la réunion la plus convenable pour discuter ces questions, s'il y a lieu. Là je me ferai un devoir de répondre à toutes les objections que l'on voudra bien me faire, et que je sollicite même vivement dans l'intérêt de la vérité.

P.S. Cette exposition a été accompagnée de démonstrations graphiques qui ne peuvent être reproduites ici, mais que je publierai ailleurs. — They are placed at the end of the Atlas accompanying "Études sur les glaciers," published three years later, as plates 15, 16, 17, and 18, Neuchâtel, 1840.

CHAPTER VI.

1836–1837 (*continued*) and 1838.

DISCUSSION RAISED BY AGASSIZ'S DISCOURSE AT NEUCHÂTEL — AGASSIZ'S GREAT REPUTATION AT THE EARLY AGE OF THIRTY YEARS — DEATH OF HIS FATHER — LAURILLARD THE ASSISTANT OF CUVIER — THE ESTABLISHMENT OF HERCULE NICOLET'S LITHOGRAPHY AT NEUCHÂTEL — DR. VOGT OF BERNE SENDS AGASSIZ EDWARD DESOR AS A SECRETARY — OFFER OF A CHAIR AT THE ACADEMIES OF GENEVA AND LAUSANNE — FIRST VISIT TO THE BERNESE ALPS — TWO LETTERS TO JULES THURMANN — A VISIT TO CHAMOUNIX — THE MEETING OF THE GEOLOGICAL SOCIETY OF FRANCE AT PORRENTRUY — FIRST USE OF LITHOCHROMY FOR THE PLATES OF FOSSIL FISHES — THE GEOLOGIST ARMAND GRESSLY — AGASSIZ CREATED A "BOURGEOIS" OF NEUCHÂTEL — ORGANIZATION OF AN ACADEMY AT NEUCHÂTEL.

THE general impression, after the address was delivered, was astonishment mingled with much incredulity. It was like a pistol shot fired into the midst of the assembly. The majority were at first disagreeably impressed by this disturbance of the peace. Von Buch particularly was horrified, and with his hands raised towards the sky, and his head bowed to the distant Bernese Alps, exclaimed: —

"*O Sancte de Saussure, ora pro nobis!*"

In general, almost all the practical stratigraphists present were either opposed to it or indifferent. Even de Charpentier was not gratified to see his glacial question mixed up with rather uncalled-for biological problems, the connection of which with the glacial age was more than problematic.

The first part of the address presented in a clear way all the facts first observed by Venetz and de Charpentier, with additional observations made by Agassiz on the Jura in the vicinity of Orbe, Neuchâtel, and Bienne. The only opinion expressed by Agassiz which was opposed to de Charpentier's glacial theory, that the ice covering all the country as far as the Jura did not come from glaciers of the Alps, was an error on his part. The second part presented by Agassiz, a combination of his ideas with those of Schimper, was fully as erroneous as the theory of water and mud currents defended by de Luc, von Buch, and Élie de Beaumont. It is not surprising that de Charpentier shook his head and was sorry to see his glacial theory used as a vehicle for such biological dreams and fantastic explanations of the "rôle" played by the upheaval of the Alps. The only rational and just conception presented in the second part is, that immense masses of ice covered the earth wherever boulders and polished rocks exist, and that the earth was covered by ice at least from the north pole to the Mediterranean and Caspian seas; in a word, that there was an "Ice-age," or "Eiszeit," according to the name coined by Schimper.

The idea of an Ice-age was a stroke of genius due to Agassiz;[1] Schimper tried to explain it by means of

[1] Many years after, when the question of an Ice-age had been recognized as settled according to the views of Agassiz, I received a letter from

biological phenomena, which according to his views were the causes of the fall of temperature (*la chute de la température*). Schimper exhibits a curious combination of a dreaming philosophy and mathematical spirit with a great deal of poetical inspiration, — a most attractive man. From the first he made use of mathematical drawings in his explanations of the morphology and phyllotaxy of plants; and during his stay at Neuchâtel in 1837, he constructed, with the help of Agassiz, a synoptical table, showing the disposition, the history, and classification of the animal kingdom, which has since been published under the title of "Crust of the Earth as related to Zoölogy," as a frontispiece to "Principles of Zoölogy," by Agassiz and Gould, Boston, 1848. Shortly after, during the same year, Schimper constructed a table showing the different systems of upheaval, as imagined by Élie de Beaumont, by means of concentric circles, with a wheel in the centre showing the directions. Applying his mathematical bent to the fall of the temperature due, according to him, to the complete extinction of life at the end of each geological period, he drew the little figure which was inserted by Agassiz in his address.

Two features characteristic of the style of this celebrated discourse must occur with force to any

him, dated Cambridge, March 13, 1868, in which he said: "Ce n'était pas petite chose de se poser en adversaire de Léopold de Buch, en 1837, et d'avoir conquis sur ce sujet l'assentiment de tous les géologues, à l'exception d'Élie de Beaumont; car l'an dernier Murchison lui-même m'écrivait qu'il se rendait enfin à l'évidence. Vous savez que la part de Charpentier se réduit à avoir démontré la grande extension du glacier du Rhône. C'est moi qui ai posé la question d'une époque glaciaire et qui l'ai fait prévaloir."

reader who is a French scholar. First, it is astonishing to see so great a number of words italicized. In no one of his papers, before or since, did Agassiz use this mode of attracting attention to special points. It shows how excited he was, and how desirous to impress on his listeners and readers several points, considered by him of paramount importance in the glacial question. As a rule, Agassiz shunned such a way of securing attention. He was a good writer, and made excellent use of French, which remained his favourite language until the end of his life. However, it is easy to detect in this address of Neuchâtel a certain number of Germanisms, due to his long residences in southern Germany.

Discussions of great earnestness followed, in which all the naturalists present joined; and although Agassiz displayed a rare talent for exposition, he succeeded only in attracting attention to the practical part of his address. With his keen eyes, he immediately perceived the bad impression made by his theoretical views, and if he did not drop them at once, it was only because it was so hard for him to admit a mistake; having once proclaimed his views and opinions on any subject, he was always most persistent in maintaining them. However, in this case he recalls his theory only once, at the end of his volume "Etudes sur les glaciers," p. 328, 1840, and never mentions it again in any of his papers or addresses.

Elie de Beaumont, who arrived the day after the meeting was over, joined von Buch in his opposition, and the two, with their Italian friend, de Collegno, were much excited and painfully affected. Von Buch, who was before very favourable to Agassiz, became an opponent, and there is no doubt that Agassiz's very fair prospect of an offer of a professorship at the Berlin University was absolutely ruined from that day.

The great value of Agassiz's address lies in his more graphic description of the action of glaciers on rocks, than that given before by de Charpentier, in his paper of 1834, and in the idea of the universality of the glacial action over half a hemisphere. Besides, it drew attention more vividly to the question, and in a way which obliged every one opposed to the view of glacial action to give his reasons.

There is no other example of such a rapid rise to great scientific reputation as Agassiz enjoyed in his thirtieth year. At the age of twenty-one, when he was still a student, he laid the foundation by his publication in 1828 of Spix's Brazilian fishes; and the first numbers or "livraisons" of his "Fossil Fishes" attracted the attention of naturalists the world over. Everything he published from 1828 to 1837 is remarkable, showing a rare power of description and classification, and a facility in handling the most difficult problem of natural history. His memoirs are entirely his own work, except the illustrations; and any one who reads them will see a difference between them and similar work produced after 1837. His power of classifying fossils and his success in reducing to order thousands of specimens of fishes, a great many of which were perfect puzzles to every one, were simply marvellous; and he worked at his herculean task as no man but a man of genius could have done.

Up to that time, he had worked entirely alone. The only collaboration he had ever had was in his Neuchâtel address before the Swiss naturalists, when he combined, as he said, his views with those of Karl Schimper, on the explanation of the great ice covering, which, according to his view, had extended from the north pole as far at least as the Mediterranean Sea. It was not a success, as he had occasion at once to see before the meeting was adjourned; for "Schimperizing"—as it was familiarly called among Agassiz's friends—was anything but congenial to his audience. It is true that he abandoned, little by little, all the ideas put forward so boldly and rashly, retaining only the word "Ice-period" (*Eisnetz*); and he returned quietly to the teaching he had received so liberally from de Charpentier and Venetz. But the difficulties which arose from this collaboration, and which broke out soon after, as we shall see, were a hint that collaboration was not suited to him, and a warning to him to be on his guard in future against scientific help and associates. Instead of heeding the warning, Agassiz, on the contrary, from that time until almost the end of his life, accepted collaboration of some sort, and entered into a succession of very serious difficulties, from which he was never able entirely to extricate himself, falling from one into another, and suffering greatly through his own fault.

It is pleasant to say that until 1837 Agassiz had really committed no fault of any consequence. At the early age of thirty years he had attained the zenith of his reputation, entirely by his own exertion and his unaided works. The address of 1837, on the glacial question, may be considered as the climax of his scientific life, as far as originality of research is concerned: it was his apogee. It is not that Agassiz's publications since that time are devoid of originality; not by any means. But after 1837 he always made too much use of others in the work of writing and too often of observation; and it is easy to detect the lack of unity, and at the same time the inequality of value in all his publications after 1837. To be sure, Agassiz published a great deal more after 1837 than he did before, but the quantity did not compensate for the quality.

His good father — a true, practical, and business man — died a few weeks after the meeting of the Swiss naturalists at Neuchâtel. He much enjoyed seeing his son, still so young, the president of an assembly of savants collected not only from all the cantons of Switzerland, but even from Berlin, Paris, Strasbourg, and Frankfort. Rodolphe Benjamin Louis Agassiz was born the 3d of March, 1776, and died on the 6th of September, 1837, at Concise, in the parsonage of that beautiful village, at the age of sixty-one years. We may say of him, what we said previously of Cuvier: had he lived ten years longer, it would have been to the advantage of Louis, who so much needed good advice and restraint in his already too great expenses.

During his stay in Paris, in 1832, Agassiz was the witness of the great help afforded to Cuvier by his principal assistant naturalist, M. Charles L. Laurillard. In the laboratory, in his library, or in his cabinet, Cuvier always found everything in perfect order, and ready for the special work he was engaged in. Laurillard, born at Montbéliard, like Cuvier, possessed a great heart, a rare modesty, profound knowledge of many questions of natural history, was devoted body and soul to his great master, and was completely devoid of any ambition, except to receive and always deserve the approbation of Cuvier. Ever since that time, Agassiz's ambition had been to get, as soon as his means would allow it, his own Laurillard. He tried again and again, and always failed. It is true that men like Laurillard are very rare; but Agassiz never possessed the art of properly managing his assistants; an art which Cuvier always had. Cuvier treated Laurillard with dignity, never with familiarity, much less in a spirit of comradery and companionship. From the first day of the arrival of Laurillard in the laboratory of Cuvier, he received a regular salary. He often accompanied Cuvier in his journeys; but he had the great tact to remain in his subordinate position of assistant, taking care to keep himself always in the background.

With Agassiz it was very different; he never knew how to keep his assistants at a distance. They very soon became intimate with him, or were allowed privileges not proper to their subordinate position. In addition, the question of compensation was a constant difficulty, either through the lack of complete understanding, or through the small amount of the salaries. In a word, Agassiz was a very bad manager of men, while Cuvier, on the contrary, was a capital and rare director of everything relating to scientific work and scientific assistants. Years after the death of Cuvier, I have heard Laurillard speak of him with the same respect as if Cuvier had been in the room. With Agassiz, all his assistants became so familiar and so much on an equality, as to raise the question who was truly the master and director.

Finding constant difficulties in regard to the execution and correction of the plates for his "Poissons fossiles," it was natural that Agassiz should desire to have a good lithography established at Neuchâtel. But such an establishment in so small a town as Neuchâtel then was, was a very hazardous undertaking; for it was certain from the beginning that the only customer of any consequence for a great lithography would be Agassiz, and, with his small salary, although raised from $400 to $600, it was almost an act of folly to establish a lithography; more especially since he was also obliged to pay for all his printing. The man chosen was a Neuchâtelois from La-Chaux-de-Fond, named Hercule Nicolet; a good lithographer, or artist rather, but as devoid of business capacities as Agassiz. The lithography was established at the end of 1836, aux Sablons, above the city of Neuchâtel, just at the place where the railroad station now stands. The establishment soon increased, about twenty persons being employed there, and turned out perfect work. But, from the beginning, it was evident that other publications with plates, besides the "Poissons fossiles," the "Echinodermes," and the "Poissons d'eau douce," must be undertaken to keep such a large establishment in work. And Agassiz, unpractical as he was, resolved to publish a German and French edition of "Sowerby's Mineral Conchology of

Great Britain," a very expensive work. The first part, or "livraison" is entitled, "Sowerby-Mineral-Conchologie Grossbritanniens; deutsche Bearbeitung, herausgegeben von H. Nicolet, durchgesehen von Dr. Agassiz," and was offered by the editor to the library of the Helvetic Society of Naturalists, at the meeting of July, 1837, at Neuchâtel.

Perceiving that he had too many irons in the fire, Agassiz longed for a secretary; and, in a visit to Berne, during the fall of 1837, he asked Dr. Vogt, the father of Karl Vogt, if he knew any young man able to write well, with some knowledge of natural history, and acquainted with the French language, because his publications must be in that language. And he added, "If you can find for me somebody of that sort, Papa Vogt, I shall bless the day which has brought me here." Karl Vogt, then a young university student, who was present at the visit of Agassiz, and has recalled the whole conversation in his biography of Edward Desor, says, "Desor had gone to Hofwyl to offer his services at the great educational establishment of von Fellenberg,[1] with the hope of being accepted. But after two days passed there, he returned to Berne absolutely crushed by his failure to obtain a position, having received the discouraging answer from Herr Fellenberg

[1] The Hofwyl College, placed by von Fellenberg in his chateau, was a philanthropic institution created as a normal school of agriculture and a model farm. There was besides a great school for secondary and superior education. It was very expensive to von Fellenberg, who was obliged to appoint too many professors of an inferior quality and who were poorly paid, and who did not stay long. They were recruited mainly in Germany among students who had just left universities.

that all the places he was able to dispose of had been already filled." The house of Dr. Vogt was a sort of refuge, always open to all German political refugees, as Desor was. At supper Dr. Vogt said, "What do you think, Desor, of going to-morrow to Neuchâtel where Agassiz is now; he wants a secretary. It seems to me that it may be a good thing for you. I will give you a few lines of introduction to him." At those words Desor jumped with joy, and, next morning, started on foot. He arrived a day later at Neuchâtel, and with his traveller's stick in his hand, a cap on his head, a gray blouse on his back, and very few pennies in his pocket, called at Agassiz's apartment and delivered the letter of introduction and recommendation of Dr. Vogt. He was accepted by Agassiz, but without any regular pay. Agassiz gave him a room in his own apartment, and paid his board at Professor Ladame's "table de pension," and as to pecuniary remuneration, it was simply understood that when he wanted money, if Agassiz had any, he would give him some; if Agassiz had none, he would have to wait until Agassiz's purse was replenished in some way. As Karl Vogt says, "When Agassiz had money, he gave what was wanted," — a singularly unbusiness-like arrangement.

P. J. Édouard Desor, born February, 1811, near Frankfort, was a law student at the University of Heidelberg, when a revolution took place in southern Germany, about 1832, in which he participated, like many other students; and he was obliged to fly to France for safety, and went to Paris, where he lived four years in poverty, giving a few lessons as a private teacher, and helping in the translation into French, and in collaboration with E. Duret, of one volume of Karl Ritter's Geography on Africa, and of a small memoir by A. de Klipstein and J. J. Kaup on the *Dinotherium giganteum*. His knowledge of natural history was very limited, and consisted only of what any student who followed lectures at Heidelberg and Paris would pick up. He had studied law, and had received no proper education to become a naturalist. He offered himself at the Hofwyl Institut, near Berthoud, directed by the celebrated Fellenberg, as a teacher of modern languages, more especially of French, for which he had fitted himself during his four years' stay in Paris. Agassiz saw at once that his natural history knowledge was most elementary; but as he was able to make good translations into French and German, and was intelligent and ready to undertake anything to get his living, Agassiz engaged him.

It must not be supposed that Desor was taken by Agassiz as collaborator and assistant in natural history. He was taken only as a secretary; for, as we have said, until then the natural sciences were almost completely unknown to him. His only duties at first were to write letters under Agassiz's dictation, to keep the accounts, to oversee what was going on at the lithography and at the printing-press. During the first two years of his stay at Neuchâtel, he took only the scientific title of *geographer*. But he followed Agassiz's public lectures, and quickly apprehended everything said by Agassiz, learning natural history with great facility. He had a good memory, and was a hard worker, — "infatigable," as Vogt says. In fact, Desor entered Agassiz's house, with the smallest possible amount of natural history knowledge; and in two years he became a tolerably good assistant in natural history, being the best pupil Agassiz had during his stay in Europe.

It is important to remark that at the time of Desor's arrival at Neuchâtel as Agassiz's secretary, nine parts, or "livraisons," of the eighteen composing the whole work of "Les Poissons fossiles," had already been issued; that is, half of the work had been published. The tenth "livraison" was on the point of being distributed, and was officially issued at the beginning of 1838. Eight plates of echinoderms, for the "Echinodermes fossiles de la Suisse," were already printed, as well as a certain number of plates of "Trigonia" and "Mya."

As soon as he was established in Agassiz's house, Desor was put at work on the translation into German and into French of Sowerby's great work on the fossils of Great Britain, and afterward at the translation into German of Buckland's Bridgewater Treatise on Geology, all of which were almost useless, not one ever having paid the expenses of printing and lithography. If Agassiz had had millions at his disposal, it would have been very well; but even then he might have used the money with more profit to science. For if up to this time Agassiz had experienced great difficulties and stringency in money matters in keeping his two draughtsmen, and publishing his "Poissons fossiles," he had at least succeeded in keeping free of heavy debts. His new undertakings were regarded with apprehension by all his family and his best friends. But it was useless to oppose Agassiz; he would listen to nothing and to no one. Science was paramount with him; everything else was of little consequence. He was born to give great impetus to natural history; and all his life he was absolutely devoted to it. Desor saw this very quickly, and took advantage of it. Science and friends working in the same field were everything. "Agassiz et ses amis," or "Agassiz et ses compagnons de voyages," became supreme. It was an unfortunate day for the future of Agassiz when Desor entered his service. From that time until he left Neuchâtel, in 1846, during nine years, expenses increased, until a complete collapse came as the inevitable consequence. Instead of being encouraged to expend more and more, Agassiz, on the contrary, ought to have been constantly restrained, on account of his too great propensity to throw money in all directions, even when it was not absolutely necessary. It was difficult to stop him, it is true; but repeated representations, accompanied by the warnings constantly poured into his ears by all the members of his family, Alexander Braun included, and all his best friends, might have resulted in restriction instead of constant expansion.

The Academy of Lausanne, after conferring on Agassiz the title of honorary professor, offered him, in 1838, a chair of active professor. Pressure was exerted by some of Agassiz's kindred, all Vaudois, — for the Canton de Vaud is the true *patria* (fatherland) of the Agassiz, — but in vain. He had cast in his lot with Neuchâtel, and remained faithful to the place which first gave him an official position. To reward his attachment, the citizens of his adopted city wrote him a letter of thanks, announcing at the same time that his salary had been increased by 2000 francs ($400) for three years. A few weeks before the offer of the Lausanne Academy was made, Agassiz was approached by the already celebrated physician, Auguste de la Rive, on the subject of a chair at the Geneva Academy. In a letter dated May, 1838, de la Rive stated frankly how the matter stood; and that he himself and everybody at Geneva thought that Agassiz was the one indispensable man. But Agassiz was already too strongly bound by his lithographic establishment and printing works to break his connection with Neuchâtel; at least, he thought so, and declined the friendly offers of de la Rive. It was doubtless a mistake; for Geneva would have given him more support and income than he was able to get at Neuchâtel. As de la Rive told him, "at Geneva you would be a second de Saussure."

After a short journey to Paris, during July, 1838, in connection with his work on the "Poissons fossiles," and to examine more carefully than he had done before the method of the laboratory at the Jardin des Plantes for moulding fossil animals, he started for the Hassli in the Oberland of Berne, studying carefully all glacial marks round the village of Guttannen, the Handeck, at the Grimsel, and at the glacier of Rosenlaui. Agassiz took with him five persons, making a party of six, — his brother-in-law Max Braun, a mining engineer just returned from Algeria, his draughtsman Dinkel, who had returned from his three years' stay in England, his secretary Desor, and two students or amateurs of glaciers and high alpine region.

It was Agassiz's first excursion to the Bernese Alps, and everything enchanted him; from Thun to Interlaken, Meyringen, and Helleplatte, where the granite is so finely polished and striated by old glaciers that it looks like polished marble. Dinkel made an exact drawing of it, which was published afterward by Agassiz in the beautiful atlas accompanying his "Études sur les glaciers." Agassiz was particularly impressed by the Grimsel and its environs, and at this time made his first visit to the glacier of the Aar, which afterward became his great station for glacial observations. The excursion lasted only ten days, and they were again at Neuchâtel on the 24th of August.

The following letter addressed to the great Jurassic geologist, Thurmann, on the occasion of the approaching meeting of the Geological Society of France in Switzerland, is interesting; for it was written on the same day he returned from the Oberland.

NEUCHÂTEL, le 24 août, 1838.

MONSIEUR JULES THURMANN,
à Porrentruy.

Monsieur, — J'arrive en ce moment à Neuchâtel d'une tournée dans les Alpes bernoises, où j'étais allé inspecter cette partie de la série de nos glaciers, désirant remettre sur le tapis la question des roches polies, des moraines, des blocs erratiques, etc., qui est si évidente et sur laquelle la plupart de nos géologues ont si peu de faits à leur connoissance. Tout ce que j'ai énoncé précédemment sur cette grave question se trouve confirmé sur un nouveau terrain où j'ai même rencontré un collaborateur intelligent, avec lequel je n'ai pas eu de peine à m'entendre, car *il avait vu* (ce collaborateur était Arnold Guyot). C'est là une condition *sine qua non*. Je suis bien réjoui que vous avez songé à faire passer la course (de la Société Géologique) par le Landeron, là, il y a de quoi voir, tout ce qui dans la question concerne le Jura; mais malheureusement nous n'avons les Alpes qu'à l'horizon et non pas sous nos pieds pour les comparer.

Cependant j'apporterai quelques échantillons, qui suppléront du moins, aux yeux de ceux qui n'auront pas pris d'avance le parti de ne pas vouloir *voir*. Je suis décidé à ne parler que de faits, les comprendra qui pourra. À moins qu'on ne veuille pas prendre l'engagement de ne pas discuter sur des suppositions gratuites et nier pour cela, l'existence des faits que l'on pourrait aller constater dans quelques journées. J'ai trop à me plaindre de la manière dont on a traité des observations consciencieuses pour vouloir prendre part une seconde fois à un pareil scandal (cela se rapporte aux critiques injustes et assez acerbes de von Buch et Élie de Beaumont à Neuchâtel l'année précédente).

D'ailleurs soyez persuadé, Monsieur, que je me fais une fête d'aller à Porrentruy, et que je compte m'y trouver dès le 4 (septembre) au soir. Notre ami Gressly est gravement indisposé, je crains bien qu'il ne puisse pas être des nôtres (Gressly n'a pu se rendre à la réunion).

J'espère beaucoup de notre visite au Landeron pour l'examen de la question des anciens glaciers et des grandes nappes de glace anté-alpines.

Agréez, Monsieur, l'assurance de ma considération distinguée.

LS. AGASSIZ.

A preceding letter to Thurmann, dated Neuchâtel, 27 January, 1836, after Agassiz's return from England, contains the following judicious remarks: —

C'est peu de jours avant mon départ pour l'Angleterre que j'ai reçu l'intéressant envoi de fossiles que vous m'avez adressés. Maintenant je vais m'occuper de les examiner. Ces objets sont d'autant plus précieux que j'en ai vu de semblables dans les terrains jurassiques d'Angleterre, et que nous avons des termes de comparaison précis pour les gisements. J'ai regretté que mon absence, m'a privée du plaisir d'assister à la réunion de la Société géologique du Jura. société qui sera d'une grande utilité pour éclaircir les questions géologiques de notre pays. [The meeting of this society, founded by Thurmann at Neuchâtel in 1834, was at Besançon, in September, 1835, and it was during the session of Besançon that Thurmann proposed to give the name *Neocomian* to the lower part of the Cretaceous rocks.] Les questions de mode de dépositions des roches, du rôle des polypiers, des équivalents géologiques, de la succession des fossiles, de leur apparition et de leur disparition sur la terre, des soulèvements, etc., ne se présentent nulle part d'une manière aussi engageante que chez nous. Je compte assister à votre prochaine réunion [that society never met again, after the Besançon meeting of 1835]. et je me réjouis de penser que vous avez déjà donné une face nouvelle à l'étude du Jura.

Farther on he adds: —

La clef des Alpes est dans le Jura me répéte M. Voltz, et il ne paraît plus douteux à ce dernier que les assises calcaires supérieures des Alpes [then called *Alpine limestone*] sont la continuation immédiate de notre terrain crétacé, et s'il en est ainsi, on pourra paralleliser toutes les couches des Alpes avec les affleurements des différens soulèvements jurassiques.

On the day after his return from the Oberland, he again left Neuchâtel en route for Chamounix, taking with him, besides his first companions, several artists and a young doctor, making a company of ten persons. Their first halt was at Bex, the Mecca of glacialists, to visit de Charpentier; who, with his usual generous hospitality and good nature, received the whole party and showed all the glacial remains, and even the stratigraphy of the salt mines, giving most clear and important explanations. From Bex Agassiz and his party visited, on foot, the valley of the Rhone between Bex and St. Maurice, the traces of the landslide of the Dent-du-Midi, and all the places most remarkable for glacial action shown to Agassiz two years previously by Venetz and de Charpentier. Crossing from the Valais by Valorsine to the valley of Chamounix, they visited the "Glacier des Bois," Montanvert, and "la mer de glace," whence they returned by the Col de Balme to Bex, visiting on the way the "Glacier du Trient."

As soon as they had returned to Neuchâtel, after a week's absence, they started again, but this time to go to the meeting of the Geological Society of France, at Porrentruy, the 5th of September. The meeting, presided over by Thurmann, was largely attended and most important. The glacial question was debated in the most spirited way; for de Charpentier was there, and Agassiz, excited by his presence, surpassed himself in trying to convert to the new theory every one present, and gave a vivid exposition of what he had just seen in the Bernese Oberland and at Chamounix. This time he was more successful than at the meeting of Neuchâtel, the year preceding. His celebrated address of 1837 had excited the curiosity of many geologists; and some, like Captain Leblanc of Montbéliard, had gone so far as to find undeniable proofs of the existence of ancient glaciers among the Vosges Mountains, proofs which were presented before the society as new facts to be added to those of Venetz, de Charpentier, and Agassiz, observed in the Alps and the Jura. One of the first converts to the glacial theory was the celebrated d'Omalius d'Halloy, who acted during the meeting as vice-president; and Bernard Studer, already well known as the geologist best informed in regard to the Bernese Alps, as well as the Molasse of Switzerland, and until then a most stout opponent, was compelled by Agassiz's explanations and enthusiasm to moderate gradually his opposition. As he afterwards said to me, Agassiz was almost irresistible in all his explanations, having a ready answer to all objections. Agassiz presented to the society, the 6th of September, his "Observations sur les glaciers," in which, though attacked by Studer, he was sustained by de Charpentier, Hugi, Max Braun, Leblanc, Guyot, and Renoir. The paper is one of Agassiz's best; it was published first in the "Bibliothèque Universelle de Genève," Tome XX., p. 382, December, 1839, more than a year after its delivery before the French Geological Society, and again in 1840, in the "Bulletin Soc. Géol. France," Vol. IX., p. 407. In it he quotes the observations of Max Braun and A. Guyot on surfaces polished by ancient glaciers near the Lake of Thun and at Oberwald, in the upper part of the Valais. Guyot had added some new facts (*considérations*) to the observations of Agassiz; but he did not write a note of what he said after Agassiz had spoken. Agassiz's secretary was at the meeting; for we find in the list of persons present not belonging to the society "M. Desor, géographe à Neuchâtel."

In an excursion of the society from Soleure to Bienne, following the foot of the Weissenstein, and at la Neuveville, Agassiz showed numerous boulders and rocks polished by glaciers. De Charpentier agreed entirely with Agassiz, and the majority of the fellows of the Geological Society accepted the new view and the glacial theory as the only possible explanation of the phenomena.

Directly after the meeting of the Geological Society of France at Porrentruy, Agassiz left for Germany to attend the meeting of the Association of German Naturalists at Freiburg-im-Breisgau in the Grand Duchy of Baden. During the sessions, which lasted from the 18th to the 24th of September, 1838, Agassiz had occasion to repeat, with great force, all the arguments relating to glaciers, the glacial doctrine, and the existence of old glaciers in the Jura, the Vosges, and the Schwarzwald. On the 25th, accompanied by Prince Charles Lucien Bonaparte, and Professor and Mrs. Buckland, he left Freiburg for Neuchâtel.

At Neuchâtel, Agassiz had his hands more than full. The lithography he had established, under the direction of H. Nicolet, turned out splendid plates of fossil fishes, by a new process of printing in various tints on different stones; what has since been termed chromolithography or lithochromy. Nicolet had engaged at Paris a French artist of great ability, Auguste Sonrel, who managed admirably with large plates, and succeeded in printing folio plates with a remarkable uniformity of colouring, as may be seen in the atlas of the "Poissons fossiles."

The studio for moulding, under the direction of Stahl, a most skilful moulder, was actively at work making casts of the inside of shells and of echinoderms, and also of topographical reliefs of the Jura Mountains, by Gressly, to show their geological structure.

We must at this time mention an addition to the staff of employés under Agassiz. There was at the meeting of the Swiss naturalists at Neuchâtel, in 1837, a very odd kind of antediluvian or primordial man, so antiquated that he seemed as if he belonged to the Jurassic period and not to our time; very awkward, timid, extremely modest, and yet so learned in practical geology that no part of the geology and palæontology of the Jura had escaped his researches. He knew every topographical feature of the Jura, every group of strata, and almost every kind of fossil remains. With great embarrassment he presented to Agassiz a letter of introduction from Jules Thurmann, the great Jurassic geologist of the Bernese Jura, by which Agassiz was informed that the name of the young geologist before him was Armand Gressly. Gressly, in the hurry of the meeting, did not dare to take from one of his large pockets the manuscript of his "Observations géologiques sur le Jura

Soleurois," but waited until after the session was over to call at Agassiz's home and present it for publication in the "Mémoires de la Société Helvétique des Sciences Naturelles." After reading the first twenty pages, Agassiz promptly saw that it was a paper of the first order, containing a quantity not only of new materials, but also of new ideas. In it Gressly proposed the theory of *facies* or "aspect des terrains," as he called it, an expression which has been constantly used the world over to explain the different association of species of fossil form, according to the deposits in which they are buried, or more exactly according to the character of the sea-bottom on which these animals had lived and associated.

Not only was the long memoir of Gressly, with a quantity of coloured sections and panoramic geological views, accepted by the committee of publication of the Swiss Society, of which Agassiz was the president, but Gressly was also closely interrogated and, as it were, interviewed by Agassiz. Although Agassiz had already met all the leaders of geology and palæontology, and a great number of practical collectors of fossils, he had never met such a curiously original observer. Gressly possessed in a rare degree precisely what Agassiz wanted, — the ability to observe the stratigraphy and to classify the different groups of rocks of a formation. Agassiz saw at once all the service he would get from such a rare practical geologist, and he offered to purchase his collection for the young Neuchâtel Museum, just organized, and proposed to him to go into the field for fossils and bring back all he could collect, arranging the specimens by strata, cleaning them, and further to revise the practical geology of all his publications on fossils. No regular pay was to be given, for Agassiz's money was already engaged to defray more than it could reasonably provide for; but Agassiz promised to provide his lodging and board, to pay his travelling expenses, and to give him money when wanted for his personal needs, if at such times Agassiz had any. In a word, it was the same unbusiness-like arrangement which Agassiz used almost all his life, and which was a constant source of difficulty with all his assistants. With Gressly the arrangement was perfectly satisfactory, and, strange to say, he was the only one who never gave any trouble to Agassiz. But Gressly was so easily contented, so timid, and had so few wants, that he was the cheapest savant imaginable to support.

A few details will give an idea of the man and his very limited requirements. Agassiz had to pay for his lodging, which consisted of a small bedroom, poorly furnished, and which soon became a true pandemonium of the most sordid kind. He boarded when in Neuchâtel at a third-rate inn called *Le Poisson*, kept by the sister of the artist Jacques Burkhardt. When travelling — always on foot — there was even less expense; for Gressly entered the first farm on his road, and asked for food and lodging. He had already roamed all over the Swiss Jura Mountains to make the observations which had resulted in his excellent "Observations géologiques sur le Jura Soleurois," and was well known personally or by reputation by almost all the country people, who always received him kindly, giving him a place at their table and a bed to sleep in—or more exactly on; for he slept with his clothes on, even with his shoes on. The farmers liked Gressly extremely, because he not only told good stories, but also gave good advice for finding springs, digging wells, and he indicated good places for marls and clays used in agriculture, and for stone quarries. Like a child, as he was all his life, he played with the children, making cocks and boats and dancing frogs out of pieces of old almanacs or newspapers. As an example of his cheap way of travelling, he once started with a small sum of money in his pocket, then he forgot that he had any money, and remained two or even three months without spending a penny, going from farm to farm, and returned loaded with the most splendid and rare fossils. And when asked why he had stayed so long without writing,—"Why!" said he, "you forgot to give me any money, and I was obliged to do as well as I could with my friends the *paysans*, who generously gave me board and lodging as I went along; a slow process," he added, "which took much of my time." "But, Gressly, I gave you some money before you started, and I saw you, if I remember rightly, put it in that pocket," indicating the pocket. Gressly put his hand in his pocket, and brought out the gold pieces which had been there, forgotten, ever since he started two months before.

As to clothes and linen, he was even more indifferent; with the exception of two pairs of strong shoes, a knapsack to put his specimens in, and a medium-sized geological hammer, everything was of the cheapest kind. He carried no change of clothing, but added shirt upon shirt, whenever he received a new one; and instead of appearing the rather slender man that he was, he gradually assumed the appearance of a very large and bulky workman. Indeed, he was constantly taken for a quarryman or a mason.

It was Gressly who collected all the materials for Agassiz's Monograph of the *Mya* and *Trigonia*, and also the majority of species of fossil echinoderms used by Agassiz in his works on that family of sea-urchins, and he also collected almost all the Jurassic and Neocomian fossil fishes.

Agassiz was not a business man, but he had found in Gressly one even less able to care for money matters, so they lived in perfect harmony. But sickness came to Gressly, who, after rallying, ended his life prematurely, at the age of fifty-one, in an insane asylum at the Waldau, near Berne. Gressly's admiration and respect for Agassiz lasted as long as his mind was not obscured.

It has been said that Desor wrote a large part of the "Observations géologiques sur le Jura Soleurois," and that Gressly was a pupil of Agassiz; but this is altogether a mistake. The manuscript of Gressly was written during 1836 and 1837, in the library of Thurmann at Porrentruy. Thurmann was constantly asked for help, which was always readily given, and read over and corrected all the rather numerous Germanisms in the French of Gressly. When Gressly started from Porrentruy to go to Neuchâtel, he carried with him his manuscript, which was delivered into the hands of

Agassiz in July, 1837, three months before Desor came to Neuchâtel, and before the name of Desor had been heard by either Gressly or Agassiz. To be sure, Gressly did learn some palæontology during his stay with Agassiz, and felt his influence; but Gressly above all was a practical geologist and a practical palæontologist, who learned all he knew of those two sciences, as he himself told me, from Professor Voltz of Strasbourg, and more especially from Thurmann; and he always called himself the pupil of Thurmann.

In 1838, two events of interest to Agassiz happened in Neuchâtel: first, his unanimous election by the counsellors of the city as "Bourgeois de Neuchâtel," and the second, the foundation of the Academy. We read in the Manuals of the Council of Neuchâtel, kept at the City Hall, the following deliberation, April, 1838: "M. le maître bourgeois en chef rappelle les services considérables que M. Jean-Rodolphe-Louis Agassiz, originaire d'Orbe et de Bavois, canton de Vaud, professeur d'histoire naturelle, rend à la Bourgeoisie et au Pays en général par l'application qu'il fait de ses vastes connaissances à l'enseignement public et par le lustre que sa réputation universelle répand sur notre patrie et sur la ville de Neuchâtel en particulier, à laquelle il donne des preuves du plus sincère attachement, ayant refusé des offres avantageuses et réitérées de places dans les cantons voisins et dans les premières universités de l'Europe. Des mérites aussi distingués ont déterminé MM. les Quatre-Ministraux à proposer au Conseil de faire attribuer à ce savant, que déja le Gouvernement a naturalisé sujet de cet État, la qualité de *Bourgeois de Neu-*

châtel, et cela a titre gratuit. Sur quoi par délibération consultative, le Conseil a unanimement approuvé la proposition, et à la même unanimité il a confirmé cette délibération au scrutin pour être soumise à la ratification de la communauté." The title of "Bourgeois de Neuchâtel" was more than merely honorary, for it carried with it pecuniary benefits, and is very seldom conferred gratuitously.

The king of Prussia, Frederick William, by an order to his Secretary of State, dated Berlin, March 17, 1838, gave for ten years ten thousand louis to develop public instruction in Neuchâtel, and Agassiz was confirmed as professor of natural history; he did not receive his diploma, however, until 1840, for two years passed before the Academy was finally organized. Arnold Guyot was invited to deliver a course of lectures on geography, a year later, 1839–1840, and Dubois de Montperreux, the Caucase traveller, also delivered lectures on archæology. For a small town and small canton, as Neuchâtel was then, the creation of an academy was a great occurrence, and did honour both to the prince of Neuchâtel, king of Prussia, and to the City Council of Neuchâtel.

CHAPTER VII.

1839–1840.

AGASSIZ'S SCIENTIFIC ACTIVITY; THE HELP RENDERED BY HIS SECRETARY DESOR — AN INTERESTING BUSINESS LETTER TO PICTET — DISPUTE WITH EDWARD CHARLESWORTH ABOUT THE FRENCH AND GERMAN TRANSLATION OF SOWERBY'S "MINERAL CONCHOLOGY" — VISIT TO THE MONTE ROSA AND THE MATTERHORN — THE GEOLOGIST VOLTZ OF STRASBOURG — STUDER'S CONVERSION TO THE GLACIAL DOCTRINE — OLD GLACIERS IN THE VOSGES — SEARCH ON THE GLACIER OF THE AAR FOR HUGI'S OLD CABIN — KARL VOGT'S ARRIVAL AS ASSISTANT TO AGASSIZ — THE HOUSEHOLD AND LABORATORY OF AGASSIZ AT NEUCHÂTEL — THE "ECHINODERMES FOSSILES DE LA SUISSE" — "ÉTUDES SUR LES GLACIERS" — THE "ESSAI SUR LES GLACIERS," BY DE CHARPENTIER — LETTER OF AGASSIZ TO DE CHARPENTIER — THE "HÔTEL DES NEUCHÂTELOIS" ON THE AAR GLACIER — VISIT OF MRS. AGASSIZ AND ALEXANDER TO THE GLACIER — JOURNEY TO ENGLAND — THE GLACIAL THEORY IN ENGLAND — AGASSIZ'S DISCOVERY OF ANCIENT GLACIERS IN SCOTLAND, IRELAND, AND ENGLAND — LETTER TO HUMBOLDT.

THE scientific activity of Agassiz during 1839 was something unique in the history of natural history researches. His secretary, Desor, had made such progress under the direction and teaching of Agassiz, that he began to be useful in original scientific observations. With a remarkable capacity and a marvellous elasticity of mind, Desor, in less than two years, had learned enough of all the branches of natural history cultivated by Agassiz to be already helpful, not only in writing under the dictation of Agassiz, but also in using his own gifts in description of species and notes on the glacial question. As Vogt says: "Desor jusqu'à son arrivée chez Agassiz, ignorait et était presque complétement étranger à toutes les branches d'histoire naturelle. Infatigable au travail, Desor était en même temps un compagnon aimable et dévoué, ayant toujours le mot pour rire et maniant avec bonhommie la plaisanterie et même l'ironie gracieuse."[1]

The notes to be used in preparing the "Études sur les glaciers" were put in order by Desor; in addition, he corrected all the proofs of the "Poissons fossiles," the "Echinodermes de la Suisse," the "Mémoire sur les Trigonies," and the "Observations géologiques sur le Jura Soleurois," and began work on the "Catalogue of all Books, Tracts, and Memoirs on Zoölogy and Geology," the "Catalogus systematicus ectyporum Echinodermatum fossilium Musei neocomensis," and finally at the "Nomenclator Zoologicus."

Agassiz gave all his assistants so much to do that it was impossible to keep pace with his eager desire and ardour for scientific publication. When we remember that it was in a small town of six thousand inhabitants that such publications were all started simultaneously by the invincible will of one man, and that all these great undertakings required not only steady and hard work but also time and money, — for Agassiz from that day published everything, with very few exceptions, "aux frais de l'auteur," — it is almost incredible. We have no exam-

[1] "Discours à l'Institut National Genevois," le 23 Mai, 1882.

ple of such impulse given to natural history anywhere, even in such great scientific centres as Paris or London. His generous spirit can be understood by reading the following extract from a letter to his friend, Jules Pictet de la Rive, dated Neuchâtel, March 10, 1839: —

Je suis également bien réjoui de pouvoir vous montrer que quoiqu'éditeur *forcé* de mes publications, c'est uniquement le désir d'être utile qui me guide vis-à-vis de mes collègues qui désirent acquérir mes ouvrages. Pour les Poissons fossiles, je vous les céderai volontiers au tiers au-dessous du prix que les libraires y mettent, c'est-à-dire à 24 francs la livraison, au lieu de 36, ce qui est à peu près le prix auquel elle me revient. Veuillez dès lors me faire savoir si je dois vous en adresser un exemplaire. Dès que j'aurai calculé exactement le coût des "Poissons d'eau douce," je vous ferai savoir aussi qu'elle remise je pourrai vous faire sur cet ouvrage. Il va sans dire que ce n'est qu'aux savants, qui me demandent mes livres pour eux-mêmes que je peux et que je veux faire le sacrifice de toutes les peines que reclament des publications de ce genre.

Generosity in this case was certainly not well placed; for Pictet was a well-to-do man in a pecuniary position far superior to Agassiz's, and might easily have afforded to subscribe at the full price. But Agassiz did not know how to discriminate between those who deserved to be helped and those whose means were such that a subscription to a costly work was not a "sacrifice," but simply a scientific duty.

About this time occurred, for the first time, a disagreeable difficulty which confronted Agassiz more than once during his life. Without asking permission, or even making his intention known, he had begun a French and a German translation of the "Mineral Conchology of

Great Britain." The Sowerby brothers, who were the authors and publishers of this costly work, thought the proceeding a little too high-handed; and the editor of "The Magazine of Natural History," Edward Charlesworth, published a rather sharp article in the May number (Vol. III., p. 254, London, 1839), in which he calls it a "piracy upon the literary production of English naturalists," and adds, "Agassiz has met with the most cordial support on all sides, and in various ways, from the cultivators of science in this country; and, although it may appear harsh thus to express ourselves, we do not hesitate openly to declare our conviction that in editing a transcript in the French language of the 'Mineral Conchology of Great Britain,' its author cannot be said to have really promoted the objects of science, still less to have added to his own reputation."

Agassiz promptly answered, in an autograph letter addressed to all his correspondents and subscribers, and reproduced in French and also in English in No. 31 of the "Magazine of Natural History" (Vol. III., p. 356), entitled, "Lettre écrite par M. Ls. Agassiz à M. Ed. Charlesworth, en réponse à une article inséré dans le No. 29 du 'Magazine of Natural History.'" In this Agassiz says, "The assertions and insinuations of the article are altogether malicious and without foundation. . . . The knowledge which I possess of the most important European scientific publications has assured me that a French or German edition of the work, published at lower price (one-fourth the cost of the original work), would be rendering a real service to science, without in any way proving injurious to the original edition, for which the principal demand is in England. Would it then not be unfair to represent such a publication as a systematic piracy; as though translations of scientific works were not being made every day with the consent of the authors?" Yes; but unfortunately Agassiz had failed to get that consent from Sowerby's sons, the collaborators and finishers of the "Mineral Conchology." There lay the mistake. Agassiz adds: "I affirm that the insinuation of my having entered upon this undertaking with a view to pecuniary emolument, to be altogether unfounded. On the contrary, only three hundred copies have been struck off, and I agreed with the editor, as the price of my participation in it, that the work should not be sold at a sum above that necessary to cover the expense of its publication." In regard to his own "Poissons fossiles," he says, "I shall esteem myself fortunate to see the work translated in whatever shape it may appear." Charlesworth rejoined, reiterating all his previous criticisms, and adding others; and finally, James De Carle Sowerby wrote a letter, the 27th July, 1839, also inserted in the "Magazine of Natural History," Vol. III., p. 418, in which he approved the strictness of Charlesworth, and suggested that some protection be afforded, at least by their *brother authors*, to those who make original and costly publications. It seems from his letter that the "sale of the 'Mineral Conchology' has only been about four hundred copies, above one-fourth of which number have been sent abroad. The encouragement, therefore, for carrying on the work has hitherto been not very great."

Agassiz, from the start of his lithographic establishment, under the direction of H. Nicolet, was very anxious to procure works sufficient to keep it running all the time, without too great pecuniary loss. His intention was good, and he began the two translations in French and German with the hope of helping that numerous class of observers of limited means on the continent of Europe, to whom Sowerby's original English edition was inaccessible on account of its great cost. I myself saw a few years after, when Agassiz's translations were hardly finished, how useful they were to French and German geologists, and they really helped the progress of science in Central Europe. The only error, and it was inexcusable, was his undertaking the work without having previously obtained permission from the two sons of Sowerby, who wrote the principal part of the text, and finally engraved also the plates, after the death of their father in 1822. Pecuniarily the enterprise was a great loss to Agassiz, for after a few years, Nicolet failed, and Agassiz had to take the whole business into his own hands. Neither of the translations sold well, and other more important works on the palæontology of France and Germany soon appeared and blocked the way. Among these were the "Paléontologie française," by Alcide d'Orbigny, and the "Petrefactenkunde von Deutsland," by Quenstedt, which not only attracted attention, but from the start were paying works, — a mark of success which was never to be granted to Agassiz during his stay in Europe. Afterwards Agassiz very seldom referred to these two translations; it was a painful subject, and he confessed that it was an error which cost him time, money, and, what is of more value, reputation among some of the English naturalists. However, he retained the friendship of all the leaders in England, and it is refreshing to read such remarks as the following: "His [Agassiz's] knowledge of natural history surprises me the more I know of him, and he has that love of imparting it, and that power of doing it with clearness, which makes one feel one is getting on, and that one has caught his enthusiasm" (Life of Charles Lyell, Vol. I., p. 457). "We are great friends," Edward Forbes wrote, after the conclusion of the meeting of the British Association at Glasgow, "and were together all the Association week. I expect him here on the 21st October; he is to work over my species with me, so as to avoid useless synonyms. . . . We worked over the synonyms, freely telling all he [Agassiz] knew, and confessing all he did not know. . . . He also gave in to my classification of the Echinodermata, admitting the Ophiuridæ as a group equivalent to the starfishes, and granting that the Sipunculidæ are Radiata" (Memoir of Edward Forbes, pp. 263, 264).

At the beginning of August, 1839, Agassiz went to the annual meeting of the Helvetic Society of Natural Sciences at Berne, where discussion on the glacial question continued to attract attention. Studer, who presided over the society that year, proposed to Agassiz to go with him to see the glaciers of Monte Rosa and the Matterhorn in Valais. The party, composed of seven persons, six naturalists and an artist, Bettannier, started from Berne the 9th of August, 1839, passing by Kandersteg to see its beautiful old moraine, already celebrated, through the agency of Professor G. Bischoff of Bonn, who had announced its peculiarity, and by the Gemmi Pass to the bath of Louèche. Here they met the great geologist, Professor Voltz of Strasbourg, and a most delightful evening was passed in his company. With such savants as Agassiz, Studer, Lardy of Lausanne, Nicolet of La Chaux-de-Fonds, and Voltz, it is easy to imagine the interest of the geological subjects which they debated until almost daybreak. Studer and Lardy had been at work for several years on the geology of the Grand St. Bernard, and other Valaisan localities of the vicinity, and Voltz, who was second only to Alexandre Brongniart, the founder of correlation of strata, by means of the fossils, had worked hard at the age of several groups of rather puzzling Alpine strata, more especially the enigmatic "Poudingue de Valorsine," and his works on the Vosges, the Alsace, and the Jura Mountains laid the basis of all that has been done there since. He was the teacher of such great geologists as Thirria, Thurmann, and Gressly, and the creator of the Palæontological Museum at Strasbourg—at that time one of the richest in Central Europe. He was, beside, a most interesting talker, full of all the socialistic theories of the time, Saint Simonian and Phalansterian, as well as an ardent republican and a friend of all political refugees, from whatever nation they came. Alas! it was for Voltz one of his last opportunities to meet the geological friends who were so congenial to him, for he died a few months after,[1] regretted by all who had the

[1] Philippe Louis Voltz, born in Strasbourg the 14th August, 1784; died in Paris the 15th January, 1840.

good fortune to know him, and by none more than by Agassiz.

From Louèche to Zermatt, "roches moutonnées" and polished boulders and moraines were met in abundance, more especially near Zermatt. At that time there was no hotel of any kind at Zermatt, and the party found lodging and board at the house of the physician of the St. Nicolas valley. Tourists had not yet discovered Zermatt, and with the exception of a few botanists and zoölogists, no one ever came to these remote parts of the Valaisan Alps. When on the Riffel, Studer, who until then had opposed the glacial theory and had explained every erratic phenomenon by mud currents, was at last convinced; his only remaining objection, after admitting ancient glaciers, being that he feared the consequences. Seeing a vertical wall of serpentine finely polished, he asked the guide to what that phenomenon was due. The guide, who had not the smallest interest in the glacial question, answered with great *naïveté*, that in the country (le pays) everybody thought that it was made by the glacier, adding: "It is true that no inhabitant of the village remembers to have seen the glacier in this place, but it was there formerly, for it is always in this way that the glaciers wear away the rocks." With great honesty Bernard Studer, who had been one of the stoutest opponents of the views of Venetz, de Charpentier, and Agassiz, confessed his errors in a "Notice sur quelques phénomènes de l'époque diluvienne" ("Bull. Soc. Géol. France," Vol. XI., p. 49. Meeting of the 2d December, 1839, Paris).

This conversion of such a prominent alpine geologist induced many other Swiss geologists, who, until then, had hesitated to adopt the glacial theory as proposed by Venetz and de Charpentier, — a theory which was extended by Agassiz to embrace almost the whole of the Northern Hemisphere. It was a great gain, due mainly to Agassiz; and from that day no more serious objections were made in Switzerland.

Curiously enough, directly after the reading of Studer's paper, Renoir of Belfort published a most important paper on the glaciers of the southern part of the Vosges. In it he declared that when Captain Le Blanc of the French Engineer Corps, at the meeting of the society at Porrentruy in 1838, announced the existence of old moraines in the Vosges, he disbelieved him; but started at once for the valley of St. Amarin as soon as the meeting was over, and to his astonishment found there proofs of all the glacial phenomena as established by Venetz, de Charpentier, and Agassiz.[1]

The proofs given by Professor Renoir, as well as the argument advanced by Captain Le Blanc, left no further doubt as to the existence of glaciers during the Quaternary period in the Vosges, and Professor Fargeaud of Strasbourg had extended his observations on ancient glaciers even to the Black Forest of Baden, and to the Pyrenees. So promptly did Agassiz's prophecy in the address at Neuchâtel in 1837 receive confirma-

[1] Note sur les glaciers qui ont recouvert anciennement la partie méridionale de la chaine des Vosges ("Bull. Soc. Géol. France," Vol. XI., p. 53, Paris).

tion beyond the Alps. After leaving Zermatt, and on an excursion to Mont Cervin, Agassiz and his party visited the glacier of Aletsch, the greatest of all the Alps, with its Merjelen (Méril) lake, unique in Switzerland; then the glacier of the Rhone, and afterward the Grimsel again. Agassiz, desirous to see the place where the monk Hugi of Soleure, some years previously, had established a cabin on the glacier of the Aar, took a guide at the Grimsel and ascended the valley of the Ober-Aar. After a rather exhausting walk over the glacier for three hours, the guide showed a well-preserved cabin on the median moraine close by an enormous granite boulder. In this they found a bottle containing several papers, one of which informed them that, in 1827, Hugi constructed a dry-walled cabin with a floor of hay, and from a second paper, also written by Hugi, they learned that he had visited his cabin again the 22d of August, 1836, and found that it had descended the glacier 2028 feet since it was built in 1827. Agassiz was much impressed by this discovery of Hugi's cabin and its motion, and he then resolved to return the next year and imitate Hugi in order to continue his researches on glaciers.

During the excursion Joseph Bettannier, who was a good landscape-artist, made several very exact drawings of the glaciers round Zermatt, Monte Rosa, Viesch, Finelen, Aletsch with its lake, St. Théodule, and Aar with Hugi's cabin, to be used for an atlas to accompany the "Études sur les glaciers."

Agassiz returned to Neuchâtel at the end of August. Soon after, an important change was made in the household involving an important addition to it. Young Karl Vogt, just graduated doctor from a German university, arrived on the last day of August, as had been agreed two years previously in October, 1837, when Agassiz was visiting his father at Berne. Karl was to help Agassiz in his publication and researches touching fossil and living fishes, and new arrangements became necessary to meet the increase of expenses. Never practical, and becoming more and more accustomed to gather round him as many assistants and social companions as he could, Agassiz could find no better way to diminish his expenses than to give Desor and Vogt their board at his own table; Desor already had a room in the house, and another near by was taken for Vogt. In this way Vogt and Desor became members of his family, their board and lodging being entirely at Agassiz's expense. As to salary, nothing was stipulated; but when they wanted money they had to ask for it, and if Agassiz had any, which was more and more rare, he gave them some. At first the new arrangement worked very well. Agassiz had company at his meals, which was always a great pleasure to him, for he was delighted to be surrounded by brilliant and intelligent, especially scientific people. Agassiz's mother, who was visiting him at this time while his wife was in Carlsruhe, was a capital housekeeper, with much dignity of manner, and accustomed to keep every one in his place without allowing the slightest encroachment or too much familiarity.

Karl Vogt in his twenties was a character seldom

met with. Tall and very corpulent for his age, his movements were rather heavy and somewhat awkward. He was inclined to see the comical side of everything, and his remarks were all tinged with ridicule. As soon as he entered Neuchâtel, he was saluted by the nickname "Le Moutz" (*Mutz* in the dialect of Berne), a popular character well known all over Switzerland, and personifying the Bernese bear; and the name clung to him during his five years' stay at Neuchâtel.

Vogt's "bon mots" soon became proverbial, and his laughter was very infectious; so much so that he would have started a Quaker meeting into uproarious merriment, and obliged a community of Trappists to break their vows of eternal seriousness and self-control.

The reverse of the medal will appear by and by. For the present Vogt made himself as amiable and acceptable as possible. Desor, who was always imitating some one or something, adopted the same attitude, and pushed his desire to please so far, that he even accompanied Agassiz's mother to the place of worship,— quite an event for a proclaimed atheist. The German language was used exclusively at table and in the laboratory; and to a visitor Agassiz's establishment at this time of his life seemed a German settlement transferred into French Switzerland.

Vogt describes his first meeting with Gressly in the following manner: "During the fall (1839) Gressly came. Great was my astonishment when I heard

Desor apostrophize the little vagabond[1] as soon as he entered into the laboratory: 'But, Gressly, go out directly and get washed; after that I will make you acquainted with Vogt.'" Gressly with his sympathetic nature, his good temper, his many eccentricities, was the constant target and object of fun for both Desor and Vogt. What he suffered, during the six winters he passed with them, is difficult to imagine. He accepted always with a smile the most cruel practical joke, working quietly at his manuscripts, and cleaning his fossils with his tongue. As soon as the spring was begun, Gressly escaped his martyrdom in the laboratory by going into the field for eight or nine months. There the poor tramp was at least free from the sarcasms of his two persecutors. It must in justice be said that years after, when both Desor and Vogt had attained reputation and social position, they were kind to Gressly. Vogt took him as a companion during a journey to Iceland; and Desor gave him a room in his house at Neuchâtel and at his country house at Combe-Varin, until he was too ill to be taken care of outside of an asylum.

But Gressly was too much absorbed in geology to be made use of as a clerk. Desor soon found out that if Gressly was ready to be treated as a funny man, he had too much independence and was too learned to be a "saute-ruisseau," as small clerks in French notary

[1] The word is most unjust and inexact, showing that Vogt is not and never was a practical geologist. Gressly was a very steady and persistent observer, and all his explorations were always systematically carried on.

offices or bureaus are called; so he hired the young son of a peasant, named Girard, of Concise, the former parish of Agassiz's father, to be his Jack at all trades. Intelligent and desirous to become a naturalist, Charles Girard submitted to the continual and rather severe exactions of Desor; for he not only had to write under Desor's dictation, but he was constantly running between the laboratory, Agassiz's lodging, the lecture-room, the lithographic establishment, and the printing-press; besides, he was the bootblack for the whole establishment. Desor kept him very close, and punished him remorselessly by sharp reprimands, which were always accepted without a word of retort, for Desor was the head man, and not an easy one to please.

As Vogt says, during the last six years of Agassiz's life at Neuchâtel, it was a kind of scientific factory, producing more than was wanted, and glutting the market with publications, without profit to anybody. Indeed, several of the works issued might have been dispensed with, both as regards cleverness and timeliness, to say nothing of the pecuniary expense, which was always rather great, notwithstanding the cheapness of living in Neuchâtel.

However open to just criticism several of Agassiz's undertakings may be, they furnished an example of marvellous initiative and of extraordinary impulse. Every one under Agassiz's direction worked hard and well; there was a sort of rivalry as to who would do best and most. The first part of the "Description des Echinodermes fossiles de la Suisse," containing spatangoids and clypeastroids, appeared during the autumn of 1839, in the "Nouveaux Mémoires de la Société Helvétique." Until then no publication on echinoderms of such importance in regard to classification, correctness of localities, and stratigraphical position, had appeared. Gressly had had a great share in it, having found the majority of the specimens used, and having helped Agassiz in his descriptions and other details of each species. A special artist, Dickmann, was trained by Agassiz to draw Echinidæ, and the accompanying plates are excellent. The memoir was made use of at once, with great advantage, by all geologists studying the Jurassic and Neocomian series; and seldom has such an important and timely contribution to palæontology been made. The second part, "Cidarides," soon followed, in June, 1840; and the whole work is one of Agassiz's best, being remarkably clear, with excellent classification, good genera and species; all of which have been accepted and used since, in all the works on fossil echinoderms.

At the beginning of the winter Agassiz wrote a very interesting letter on the glaciers to Élie de Beaumont, asking him to communicate it to the Academy of Science. But de Beaumont was a rather unfair opponent in everything relating to the glacial question, and he did not read the letter to the Academy, as he was requested to do. As it is important, and allows every one to see the opposition at that time constantly made against the doctrine of the action of glaciers in the Alps, I give it almost in full, suppressing only local details relating to the glaciers around Zermatt and Mont Cervin. Besides, this letter had the advantage of giving, before the publication of "Études sur les glaciers," advertised to appear the following year, explanations of some of the plates prepared for the folio atlas to accompany that work.

NEUCHÂTEL, 16 décembre, 1839.

LOUIS AGASSIZ
à Élie de Beaumont.

Je venais d'emballer les premières épreuves de mes planches de glaciers pour vous les envoyer lorsque je reçus votre lettre à laquelle je m'empresse de répondre. J'espère pouvoir vous adresser d'ici au printemps le cahier complet des planches que je fais faire sur les phénomènes que présentent les glaciers; celles que je vous adresse aujourd'hui ne sont relatives qu'à la formation et à la marche des moraines et à l'action des glaciers sur le fond sur lequel ils reposent. Tout ce qui concerne la structure intime des glaciers et la plus grande extension qu'ils avaient autrefois ainsi que les moraines anciennes sera figuré plus tard. J'ai voulu rester d'abord dans des limites où je suis sûre de ne rencontrer aucune opposition. Ce sera je l'espère le meilleur moyen de préparer un acceuil favorable aux phénomènes trop contestés dont j'ai déjà parlé ailleurs. Je crois que je parviendrai à les faire adopter lorsque je serai parvenu à démontrer avec le même respect qui vous anime pour les lois générales concernant notre globe, que des oscillations de température un peu plus grandes ou un peu plus faibles ne sortent pas plus du cadre des lois invariables de la physique que des phénomènes de soulèvement poussant un îlot à fleur d'eau, ou soulevant la chaine des Alpes. D'ailleurs l'étude comparative que j'ai faite d'une part de l'effet de l'eau courante, ou de grandes masses d'eau mues par les vents, d'autre part des effets produits par le mouvement des glaciers, me permet maintenant de les distinguer à quelque distance de leur source première que je les rencontre.

Mais revenons aux glaciers tels qu'ils se présentent dans leur limites actuelles. Les Planches 1 et 2 (de l'Atlas de 32 planches des *Études sur les Glaciers*, qui a paru en 1840), que l'on peut joindre, offrent un panorama des principales sommités du Mont-Rose vues depuis le Riffelhorn. . . . Je ferai d'abord remarquer que sur la droite de la Planche No. 1, on voit distinctement une grande moraine formée autour du rocher saillant qui borde le glacier du Mont-Rose, et qui est refoulée sur le glacier principal par les glaces descendant du Lyskamm. c'est la moraine que j'appelle la grande moraine du Mont-Rose pour la distinguer d'une autre moraine moins considérable qui se forme par les éboulements, de quelques arêtes nues du Mont-Rose et qui descend à peu près sur le milieu du grand massif de glace qui sépare le Mont-Rose du Gornerhorn et qui après s'être repliée sur le milieu du grand glacier marche parallèlement avec la première. De l'angle inférieure du Gornerhorn, on voit surgir une troisième moraine séparée de la petite moraine du Mont-Rose par une série d'entonnoirs d'abord peu distincts, mais qui grandissent en face du Lyskamm et du Breithorn, pour disparaître entièrement plus bas. Je lui ai donné le nom de moraine du Gornerhorn. Enfin sur le devant de la planche on remarque une quatrième moraine qui descend du milieu de la Porte-Blanche et qui tend à se confondre avec la moraine du Gornerhorn (La Porte-Blanche est l'arrête qui domine la vallée de Macugnaga au nord du Gornerhorn). Lorsqu'on descend au pied du Riffel on remarque une cinquième grande moraine au bord du glacier, mais elle reste inaperçue sur cette planche à cause de la saillie que forment les rochers d'où le panorama est dessiné. . . .

La planche 3 représente le glacier de Zermatt au point où après avoir reçu les affluents de tous les pics il commence à descendre dans la vallée en s'engageant entre le Riffelhorn et les rochers appelés Auf-Platten. . . . La planche 4 est la continuation de la planche 3; le glacier est déjà considérablement descendu entre le Riffelhorn et Auf-Platten. Cette vue est prise vis-à-vis du Riffelhorn au bord d'une cascade qui descend du glacier de Furke en montant à Auf-Platten, tandisque les trois autres planches sont dessinées depuis le Riffelhorn. . . . En général à l'extrémité du glacier les moraines se dispersent tellement qu'il est fort difficile de les distinguer les unes des autres. On ne les reconnoit guère qu'à la nature de leurs roches. La marche de toutes ces moraines complètement distinctes dans la partie supérieure du glacier, plus ou moins confondues dans sa partie inférieure prouve que les affluents de glace qui descendent des pics supérieures comme autant de massif distincts se réunissent plus bas en un massif homogène semblable à un grand fleuve qui vers son embouchure roule d'une manière uniforme des flots longtemps distincts, mais enfin confondus dans leur marche. . . .

Les planches 6 et 7 doivent donner une idée de l'action qu'exercent les glaciers sur le fond sur lequel ils se meuvent. . . . En pénétrant sous le glacier, entre ses crevasses, à plusieurs mêtres de profondeur j'ai pu me convaincre que le poli des roches et les stries (burinées dessus les serpentines) existent uniformément sous le glacier comme sur ses flancs, et la direction des stries que j'ai observée le long du glacier depuis le pied de la Porte-Blanche jusqu'à la source de la Viège qui sort de la voute inférieure du glacier, la direction de ces stries, dis-je, qui suivent toutes les inflections du glacier, qui sont rectilignes partout où le glacier se meut en droite ligne, qui se courbent et prennent même une direction ascendantes là où le glacier passe pardessus des arêtes saillantes de rochers; cette direction ne laisse aucun doute sur la liaison qui existe entre ces stries et le glacier lui-même. On ne saurait douter non plus quand on a poursuivi ce phénomène sur une aussi grande étendue, que les grains de quartz provenant des granites triturés dans les moraines marginales ne soient l'émeri au moyen duquel le glacier en se mouvant polit et raie le fond sur lequel il marche; il me paraît impossible de supposer que ces surfaces polies et ces stries aient existé antérieurement à la formation des glaciers, et que les glaciers aient pu se mouvoir à leur surface sans les effacer. Ces surfaces polies et ces stries sont si constantes autour des glaciers, si fraîches dessous leurs masses, si bien conservées partout où les glaciers existent encore que les habitants de la contrée les ont remarquées et les attribuent au mouvement des glaces même là où le glacier a disparu. Leur direction presqu'horizontale tout le long du bassin du glacier de Zermatt, sur les flancs et parois du Riffel et d'Auf-Platten s'oppose à toute idée d'avalanche, comme cause de ces stries, car à raison de la configuration des lieux, toutes les avalanches qui pourraint se former, couperaient nécessairement à angle droit la direction des stries telle qu'on l'observe; en un mot les faits sont de telle nature dans toute l'étendue du cours des glaciers que je viens de décrire, qu'il est impossible de ne pas reconnaître que c'est le glacier qui a poli ses bords au-dessus du niveau qu'il occupe maintenant et qu'il continue à polir les rochers sur les quels il repose encore. Les faits sont si parlants que M. Studer qui a fait une fois la course du Riffel avec moi s'est rendu à l'évidence quoiqu'il eût nié jusqu'alors la liaison des surfaces polies et des stries avec les glaciers. Une autre circonstance qui parle hautement en faveur de cette liaison c'est que les surfaces polies et les stries sont d'autant moins distinctes qu'on les observe sur des surfaces abandonnées depuis plus longtemps des glaciers et où ils ont cependant existé de mémoire d'homme, comme c'est par exemple le cas au-dessous de l'extrémité actuelle du glacier de Viesch, que les régistres de la paroisse d'Aunen constatent s'être étendu jusque près du village de Viesch, c'est-à-dire une lieue plus bas que maintenant.

La vallée de Viesch est une des plus intéressantes que je connaisse pour l'étude comparative de l'action des eaux et des glaces sur le fond de leur lit; et quelque soit la cause à laquelle on attribue les surfaces polies et les stries, toujours est-il que dans chaque vallée où on les observe, elles suivent en somme la direction de la vallée, c'est-à-dire, que pour prendre des exemples precis, les stries de la vallée de Viesch s'inclinent du Nord au Sud vers le Rhône, tandisque celles qui accompagnent le glacier du Rhône sont dirigées de l'Est à l'Ouest et celles qui accompagnent le glacier de l'Aar de l'Ouest à l'Est jusqu'à l'hospice du Grimsel, puis du Sud au Nord du Grimsel à la Handeck où il est certain que ces stries existent sur les flancs du glacier de l'Aar jusqu'au niveau du col qui sépare l'Oberland bernois du Valais. Pour pouvoir attribuer ces stries à des courants il faudrait donc (abstraction faite de tous les faits que j'ai déjà cités et qui prouvent une liaison intime entre les stries et les glaciers) imaginer des courants remplissant jusqu'à les combler ces hautes vallées et dirigés l'un du Finsteraarhorn à l'Est jusqu'au Grimsel en sens inverse d'un autre courant parallèle dirigé des sommités des glaciers du Rhône vers la Mayenwand, c'est-à-dire, de l'Est à l'Ouest et se precipitant dans la vallée du Rhône pour y rencontrer un troisième courant, tout aussi puissant dirigé directement du Nord au Sud de la vallée de Viesch ; et tous ces courants devraient naître sur la crête si étroite qui sépare ces trois vallées ; car comme vous l'avez très bien observé, les surfaces polies nous prouvent que le relief du centre de l'Europe n'a subi aucun changement notable depuis qu'il est sous l'influence des causes actuelles. Or revenons à la vallée de Viesch dont la partie supérieure est occupée par un glacier et dans le fond de laquelle coule un torrent rapide dont le cours n'est pas beaucoup plus court que ne serait le grand courant auquel on voudrait attribuer les surfaces polies et les stries de cette vallée, si jamais pareil courant avait pû naître sur les crêtes du Viescherhorn, et voyons bien qu'elle influence le glacier actuel et le torrent actuel exercent sur le fond de leur lit.

Les roches au bord du glacier et sous le glacier sont polies et striées dans toute l'étendue que recouvre maintenant le glacier. Partout où l'on peut pénétrer sous la glace ou déblayer la grande moraine qui l'entoure, les stries et les surfaces polies sont fraiches et la direction des stries ne laisse aucun doute sur la cause qui les a produites, ici encore elles sont dues aux glaciers Il est vrai que le torrent qui corrode le fond de cette vallée y creuse des sillons sinueux et polit les cotés de son lit, mais ces polis effectués par l'eau ont un aspect tout différent, ils sont mats, creux, souvent même incrustés ; ce sont des coups de gouge plus ou moins allongés, limités par des arêtes saillantes ; jamais ils ne sont striés ; jamais ils ne présentent de surfaces un peu étendues, tandisque les surfaces polies par le glacier sont bosselées en relief, les parties saillantes sont surtout striées et les parties dans la roche ne font jamais saillie. Les surfaces polies qui sont encore maintenant sous le glacier dans cette vallée sont la continuation directe de celles sur lesquelles le glacier ne repose plus, mais sur lesquelles on sait qu'il a reposé jadis. Ces surfaces polies dénudées que l'on voit sur les côtés du cours du torrent sont striées dans le même sens que celles que l'on voit encore sous le glacier ; elles diffèrent complètement des surfaces corrodées par l'eau du torrent, mais elles sont identiques avec les parois de la vallée qui ont conservé leur poli. Mais comme on ne voit aucune trace analogue à celle du torrent dans la partie supérieure de la vallée, tandisque ses parois sont striées et polies à de grandes hauteurs absolument comme sous le glacier, il me parait d'une bonne logique de conclure que la cause qui a agi plus puissament autrefois que maintenant, était un glacier plus étendu et non pas un grand torrent. Je n'entrerai pas ici dans le détail des différences très notables que présentent les roches de différente nature sous l'influence des glaciers et sous celle des courants, vous l'avez sans doute déjà remarqué. Je me bornerai à dire que les serpentines de la vallée de Zermatt et du Riffelhorn présentent le plus beau poli que je connoisse; que les granites des parois du glacier de l'Aar ne le cède en rien aux serpentines là où ils n'ont pas été encore exposés à l'action de l'air, mais que l'atmosphère les rend facilement rudes au toucher; que les gneiss ne conservent guère de traces de stries et de polis, que lorsque les glaciers ont agi sur les tranches de leurs couches; que les calcaires, tout en prenant facilement un très beau poli ne le conservent pas facilement lorsqu'ils ne restent pas recouverts par le limon des moraines après avoir été polis. Cela est si vrai, que dans les Alpes ce n'est guère que sous les glaciers mêmes que les calcaires alpins conservent les traces de leurs stries; ces faits sont une nouvelle preuve bien puissante de mon assertion, que les surfaces polies et les stries sont réellement dues aux glaciers et ne peuvent point avoir été simplement conservés sous les glaciers.

Dans l'exposé de ces faits je me suis restreint aux phénomènes tels qu'ils se présentent dans les Alpes, afin de ne point réveiller les objections qui n'atteignent que leur extension dans des régions où les glaciers n'existent plus; plus tard je reviendrai sur les glaciers du Jura, lorsque l'ensemble de mes observations sera aussi concluant pour ces contrées qu'elles le sont pour les Alpes. Je dirai seulement que mes courses de cet automne m'ont fourni de nouvelles preuves de la liaison qui existe entre les blocs erratiques du Jura et les glaciers. J'y ai acquis en même temps la conviction qu'il a existé dans l'intérieur du Jura des glaciers indépendants de ceux des Alpes. Les physiciens s'arrangeront de ces faits comme ils le pourront, mais je ne crois pas qu'il y ait quelque chose de plus contraire aux lois de la physique dans les phénomènes qui nous démontrent l'existence d'une création (de faune et de flore) tropicale en Suisse, que dans ceux qui lui assignent à une autre époque un climat boréal.

En vous adressant prochainement les autres planches de mon livre je les accompagnerai de quelques observations sur la marche des glaciers, sur leur formation et sur leur structure intime. Une troisième notice sera relative aux phénomènes éloignés des Alpes qui je crois se rattachent à ceux dont je viens de vous entretenir. Avant de les publier je désirerais vous les soumettre dans leur ensemble.

Vous m'obligeriez infiniment en m'exposant d'une manière précise les objections que vous avez à faire à ces considérations. Quoique j'aie déjà fait de nombreuses observations thermométriques sur les eaux courantes et sur les petits lacs et les mares des glaciers et sur les glaces mêmes, je fais de nouveaux préparatifs pour aller étudier l'été prochain les profondeurs des glaciers en faisant des sondages pour traverser tout le massif des glaciers et pénétrer jusque dans le sol sous-jacent. Si vous aviez quelques observations à me proposer, je les ferais avec le plus grand plaisir; je serais également fort redevable à M. Arago s'il voulait bien me faire part de ses *desiderata* relativement aux glaciers. Je compte passer l'été prochain plusieurs semaines dans le cœur des Alpes.

J'ai visité cet été tous les abords de la grande mer de glace qui s'étend entre le Valais et l'Oberland afin de m'orienter préalablement, et mon intention est de la traverser dans tous les sens si le temps m'est favorable J'ai déjà pénétré par le glacier de l'Aar jusqu'au pied du Finsteraarhorn, et par le glacier d'Aletsch jusqu'au pied des Viescherhorner, derrière la Jungfrau, et passé de là au glacier de Viesch. Mon projet serait de traverser de Grindelwald au Grimsel par l'arête d'Ashchwung.

Si vous pensez que ces observations puissent intéresser l'Académie des sciences (de l'Institut de France), vous m'obligeriez en lui en communiquant succinctement le contenu. Ne jugez pas trop sévèrement mes dessins, mais pensez à la difficulté qu'il y avait à encadrer dans des dimensions données, des vues privées de toute végétation, ne représentant que des rochers nus, des glaces, et des neiges, où l'on rencontre à peine des êtres vivants, par ci par là seulement quelques Pyrrhocorax, quelques Gélinottes, rarement des marmottes, plus rarement encore des chamois et d'où les habitations des hommes se voient dans le lointain, au fond des vallées, comme dans l'abyme.

Je vais faire copier mes croquis du Jura pour vous les envoyer immédiatement. Je ne tarderai pas non plus à vous envoyer ma notice sur vos Echinodermes.

The winter of 1839–1840 was employed in writing, besides the continuation of the "Fossils Fishes," a volume on the glaciers, and two monographs on the echinoderms, and on the *Trigonia;* and Vogt translated the manuscript of the "Études sur les glaciers" into German, in order to have the French and German edition issued at the same time. The book appeared in September, 1840, with a splendid folio atlas of eighteen beautifully executed plates. In it Agassiz very frankly gives an account of his five months' companionship in 1836 with de Charpentier, who taught him the glacial doctrine, and of his returning with several of his friends: among them, Karl Schimper, Francillon of Lausanne, who became his brother-in-law, Max Braun, Dinkel, and his secretary, to visit again the classical localities first shown to him by de Charpentier. The historical part on the glaciers is very full and just to every observer who had entered the field before him. The work is dedicated to "M. Venetz, Ingénieur des Ponts et Chaussées au Canton de Vaud, et à M. J. de Charpentier, Directeur des Mines de Bex." Notwithstanding all these precautions, the work displeased Venetz, de Charpentier, and Hugi, his three predecessors in the study of the alpine glaciers of Switzerland. De Charpentier was at work on his volume "Essai sur les glaciers," which was then passing through the press, and he thought that his pupil Agassiz might have waited until he himself had given to the world his researches, before printing what he had learned from him. It was a question of politeness, which de Charpentier emphasized perhaps too strongly, for Agassiz did not intend to wound him; on the contrary, he proclaimed the priority of Venetz's and de Charpentier's discoveries. But the method used by Agassiz shows a want of courtesy in his eagerness to propagate and make known the new doctrine. A few words are necessary to explain the estrangement of friendly relations between Agassiz and de Charpentier. Agassiz, with his insatiable appetite, and his great faculty of assimilation, digested the whole doctrine of the glaciers, and made use of it, as it was almost his own. He did not want to wrong de Charpentier in any way, but he was so ardent,.so impulsive, that he appeared in the eyes of de Charpentier and his friends to be too eager in taking the wind from the sails of others. De Charpentier's manuscript was finished the 31st of October, 1840, and he received Agassiz's "Études sur les glaciers" only three days before, on the 28th of October, and thus had time only to look it over and notice it in his Introduction, pp. vii and viii. As Agassiz continued in his work to maintain his fanciful theory of transportation of boulders, by sliding over the ice-sheet, de Charpentier's objections, pp. 232–241, were timely and to the point.

The "Essai sur les glaciers" appeared a few months later, in February, 1841. Of that work the biographer of de Charpentier says: "The work will remain a classic. Unhappily the modesty of the author induced him to publish it at Lausanne, which explains why it was so little known in France, in Germany, and other countries, when, if it had been published in Paris, with a simultaneous German edition in a great city of Germany, it would have been one of the most important and at the same time popular books of the time. I cannot better express my admiration for the work than to say that it is impossible to be truly a geologist without having read and studied it" (Dr. H. Lebert, "Biography of Jean de Charpentier").

The following letter from Agassiz to de Charpentier explains the impression made on both by the publication of their two works on the glaciers and the transportation of boulders : —

NEUCHÂTEL, 28 juin, 1841.

à M. J. DE CHARPENTIER,
Directeur des Mines, à Bex.

Mon Cher Monsieur, — Après beaucoup de démarches inutiles j'ai enfin pu me procurer votre ouvrage sur les glaciers, etc. Je l'ai lu avec avidité et j'y ai trouvé beaucoup de faits d'un grand intérêt. Je me suis convaincu de nouveau que nous avons tous encore beaucoup à apprendre sur ce sujet. Je regrette une seule chose c'est que vous ayez si peu mis à profit mes observations, vous auriez pu par là compléter plusieurs points de votre travail et vous donner le mérite de fondre tout ce que l'on sait maintenant de positif sur la question des glaciers, d'harmoniser les dénominations divergentes que vous avez employées, d'établir la synonymie des vôtres avec les miennes, etc. Puisque vous n'y avez pas songé je m'en chargerai et malgré le mauvais vouloir que vous avez mis partout en me citant, vous n aurez pas trop à vous plaindre de moi, car je tiens avant tout aux progrès de la science sans acception de personnes. J'ai d'ailleurs une masse d observations nouvelles à publier, recueillies dans les montagnes des Isles Britanniques l'automne dernier et au commencement du mois de Mars de cette année sur le glacier inférieur de l'Aar que j'ai parcouru jusqu'à l'Abschwung.

L'affection que je vous ai toujours conservé me fait regretter pour vous que vous vous soyez donné le tort de critiquer des bagatelles de mes planches et de mon livre, sans citer aucun fait instructif, excepté la température du glacier. Cette réserve est tellement frappante que déjà deux de mes amis m'en ont exprimé leur étonnement. Mais cela s'oubliera j'espère.

Au revoir à Zürich si vous y allez, si non j'espère sur votre territoire un peu plus tard. Mes respects à Mademoiselle de Charpentier. Agréez l'assurance, etc., etc.

LOUIS AGASSIZ.

This letter ended the friendly relations between two unusually congenial men of genius, who ought to have remained friends, as workers in the same field and as neighbours. If left to himself, Agassiz would have bridged the chasm; but he was already too much influenced by his secretary and by some others of his collaborators, more or less interested in keeping matters embroiled. After repeatedly hearing Agassiz, and once hearing de Charpentier, I do not hesitate to say that, but for the objectionable surroundings in which Agassiz lived from 1839 until he left Switzerland, the wound would have been promptly healed and friendship renewed.

On the 5th of August, 1840, Agassiz left Neuchâtel for the Grimsel. There he took into his service two of the best Oberland guides, Jacob Leuthold and Jean Währen, the latter a mason by trade, and started at once for the lower part of the glacier of the Aar. The plan was to establish a station on the glacier itself, and for that purpose to make use of Hugi's cabin, found by Agassiz in the preceding year, in a very good state of preservation, as already reported. But to his astonishment the cabin had disappeared, and it was with some difficulty that Agassiz at last found some of the débris, two hundred feet below the place occupied by the cabin in 1839. After consultation with the guides, who gave the very practicable advice to build a cabin on the rock bordering the left side of the glacier, Agassiz, who was resolved to imitate Hugi, gave all sorts of reasons for establishing the cabin on the median moraine, and finally an enormous block of micaceous slate was selected. A part of the block projected in a sort of roof, under which a wall was built by the mason. Four porters, lent by the housekeeper of the Grimsel's hospice, to carry provisions and bedding, helped in the construction of the cabin, which was inhabited the same evening. The opening of the cabin was toward the south, and a good sketch of it has been published in the "Excursions aux Alpes," by Desor, p. 157. During the night the cabin was christened by the name "Hôtel des Neuchâtelois," which was engraved by the mason in big letters on the block, and the names of the first six occupants were a few days after added. They were Louis Agassiz, Charles Vogt, Ed. Desor, Célestin Nicolet, Henri Coulon, François de Pourtalès, the last two being students at the Neuchâtel Academy.

Observations were begun at once on every point pertaining to glaciers, including structure, motion, tables, moraines, névés, climate and meteorology, red snow, crevasses, etc. Visitors from the Grimsel came now and then; and, to the great joy of Agassiz, one day he saw wending their ways with some difficulty over the glacier to reach his "Hôtel des Neuchâtelois," his wife, her sister, Fraulein Emmy Braun, and his son Alex., the latter borne on the shoulders of the guide Jacob. That day the dinner on the glacier was particularly luxurious, fresh provisions having come with the visitors, and the pleasure of the unexpected meeting enlivened the otherwise rather rough establishment, with its numerous discomforts.

After a visit to the top of the Strahleck, the party left the "Hôtel des Neuchâtelois," after a stay of only six days, from the 10th to the 16th of August, 1840. Before returning to Neuchâtel, Agassiz traversed the Scheideck, and made observations on the glaciers of Grindelwald, of Schwartzwald, and of Rosenlaui; he visited also the upper part of the glacier of the Aar, and passed a night on the Siedelhorn.

Directly after returning to Neuchâtel, Agassiz left for England. During the meeting of the British Association for the Advancement of Science, in September, at Glasgow, he had an opportunity to see how little progress the glacial question had made among English naturalists: it was almost unknown. Buckland alone, during a protracted visit to Switzerland in 1838, and after resisting as long as he could all the facts concerning glacial action, was at last converted by Agassiz to the new theory. But his conversion had no other effect on English geologists than to bring forward a semi-caricature drawn by Thomas Sopwith, which was largely circulated as a portrait of Buckland dressed in "costume of the glaciers," and which has been reproduced since in "Memoir of Sir Roderick Murchison," by A.

Geikie, Vol. I., p. 309. The reproduction by Archibald Geikie is not, however, a complete one; all the devices and explanations written on the big roller of maps and under the scratched stones have been omitted, and even the title of the semi-caricature has been altered. It is easy to see the reasons for these suppressions and alterations. The mining engineer, Thomas Sopwith, has stated the objections made against the glacial theory in such childish and ridiculous words, that to repeat them was considered by Geikie as reflecting little credit on all those who made fun of the glacial epoch, with Murchison as their leader.[1]

[1] Here is the exact description of the semi-caricature. Buckland, equipped as a glacialist, stands on a flat bit of rock covered with scratches, with the following explanation: "The rectilinear course of these grooves corresponds with the motions of an *immense body*, the momentum of which does not allow it to change its course upon slight resistance." On the polished rocks is written: "Prodigious glacial scratches"; and in order to add to the value of the opposition made by anti-glacialists, the author has engraved, just under the last sentence, "Scratched by T. Sopwith." The title of the drawing is: "COSTUME OF THE GLACIERS." Under his right arm Buckland holds a rather large and long roller, with the inscription on it: "Maps of ancient glaciers." At his feet, on his right side, are drawn: "Specimen No. 1, scratched by a glacier thirty-three thousand three hundred and thirty-three years before the Creation"; and just below, another specimen of a "cailloux strié," marked: "Scratched by a cart-wheel on Waterloo Bridge the day before yesterday." It is now almost incredible that such objections should have been able to elicit anything more than a smile at the ignorance of plain facts.

Philip Duncan was better inspired, when he wrote in his poetic "Dialogue between Dr. Buckland and a Rocky Boulder": —

Boulder, *respondit.*

. . . .

"And many a rock, indented with sharp force,
And still-seen striæ, shows my ancient course:
And if you doubt it, go with friend Agassiz
And view the signs in Scotland and Swiss passes."

Murchison, in a letter dated Sept. 26, 1840, in speaking of the Glasgow meeting says: "Agassiz gave us a great field-day on Glaciers, and I think we shall end in having a compromise between himself and us of the floating icebergs! I spoke against the general application of his theory." This was precisely what was to be expected from the English geologists, who are always strongly disinclined to accept any new truth, if discovered by foreigners. Even the Uniformitarians, at that time already very numerous in England, with Charles Lyell as their leader, did not see the splendid opportunity to add a new crown of laurels to Uniformitarianism, or the doctrine of existing causes, and they persisted in getting entangled among masses of floating iceberg.

In company with Murchison, Agassiz visited the North of Scotland to see the Old Red Sandstone and its fishes. During the journey Agassiz found a great number of traces of ancient glaciers, and in vain showed them to Murchison, who, on the 29th of October, wrote to Sir Philip Egerton: "If you have not been frost-bitten by Buckland, you have, at all events, had plenty of friction, scratching, and polishing, before now, and next year you may give us a paper on the glacier of Wyvis and the 'moraines' on which you sport! I intend to make fight." On a question in regard to which he knew next to nothing.

However, Murchison's "fight" amounted to the old rehearsal of the floating iceberg theory and mud currents, two exploded doctrines, rather antiquated even in England after Agassiz's visit of 1840.

On the 4th of November, 1840, Agassiz read before the Geological Society of London his paper "On Glaciers, and the Evidence of their having once existed in Scotland, Ireland, and England" ("Proceed. Geol. Soc. London," Vol. III., pp. 327–332). This memoir — a masterly one — opened a new chapter in the geology of the British Isles. In the "Life of Murchison," by Archibald Geikie, we find the biographer saying (p. 309, Vol. I.) that "the remarkable series of observations by Agassiz among the glaciers of the Alps, and the extension of them to Scotland by Buckland, Lyell, and Agassiz himself," — a sentence which seems to imply that Agassiz came after Buckland and Lyell. The man who with great difficulty, and after a stout and protracted resistance, during a prolonged visit to Switzerland, in 1838, taught Buckland how to recognize traces of ancient glaciers, is represented as occupying only a third place in the discovery of the evidence of the existence of glaciers in Scotland. The truth is, that Buckland, after being converted to the new doctrine, informed Agassiz that he had noticed similar phenomena in Scotland, but had attributed them to diluvial action. He waited until Agassiz came to Scotland, and it was when in his company that Agassiz said, as they approached the castle of the Duke of Argyll, "Here we shall find our first traces of glaciers"; and surely enough, the carriage as it entered the valley rode over an ancient terminal moraine. Then, and not until then, Buckland was made sure that his indications were well based. It is important to add that Buckland did not claim any *priority*. On the contrary, he read his memoir "On the Evidences of Glaciers in Scotland and the North of England," after Agassiz's paper, and to sustain him by what he had learned in his company during the fall and afterwards. At present, to make amends for their slowness in recognizing old glaciers, the Scotch geologists, with James Geikie at their head, are claiming that they had found evidences of the existence of no less than *five glacial periods* during the Quaternary epoch.

Agassiz's three-months visit in the British Isles during the autumn of 1840 may be counted as his most successful period of happy and important discoveries, and he returned with the great satisfaction of having extended the glacial doctrine to Scotland, the North of England, and Ireland, and having first explained the complicated organization of the fossil flying-fishes of the Old Red Sandstone. The following letter to Humboldt gives an excellent résumé of his three months' exploration:—

NEUCHÂTEL, 27 déc., 1840.

à SON EXCELLENCE M. A. DE HUMBOLDT.

Mon Cher et Excellent Ami,—Je suis de retour à Neuchâtel depuis huit jours et déjà je me suis remis au travail. J'ai pris la ferme résolution de ne rien faire cet hiver que des "Poissons fossiles" et j'espère achever mon ouvrage avant l'été. Pour y parvenir je ne publierai pour le moment que les mille espèces les plus intéressantes de manière à en faire un corps d'ouvrage lié et je donnerai plus tard dans un Supplément 6 à 700 espèces que je n'ai pas encore complèment étudiées. Mon dernier voyage en Angleterre m'a fait faire des progrès réels en Ichthyologie fossile; j'ai surtout étendu mes observations sur les espèces siluriennes, dévoniennes et houillères. Les genres de l'"Old Red" sont surtout très remarquables. Le prétendu Coléoptère gigantesque de Fifeshire[1]

[1] "History of the County of Fife," by J. Anderson, 4to, Edinburgh.

est un poisson *Pterichthys!* J'ai d'autres types tout aussi extraordinaires *Coccosteus*. Ce qu'il y a de très curieux c'est que tous ces poissons ont des têtes disproportionnées, égalant, dépassant même le tronc en longueur et toujours de beaucoup plus larges, dans le style des Torpiles, mais à charpente osseuse et couverts de larges écussons émaillés On m'à communiqué en somme environ 250 espèces nouvelles. J'ai également examiné un nombre immense d'Echinodermes. Heureusement que j'ai obtenu des exemplaires de la plupart des espèces, car mon temps n'aurait pas suffi pour les décrire en détail.

Cependant je vous avouerai que ce qui m'a fait le plus de plaisir c est d'avoir découvert des traces indubitables de glaciers sur une très grande échelle. Les marques de leur présence sont si frappantes que tous les géoloques qui les ont vues sont restés convaincus du fait. Depuis que j'ai rendu compte de mes observations à la Société géologique (de Londres), les mémoires sur ce sujet se succédent. Buckland a décrit ceux qu'il a observé au centre de l'Ecosse et au Nord-Ouest de l'Angleterre; Lyell ceux du Forfarshire. Pour moi je m'étais surtout appliqué à démontrer qu'ils ont réellement existé dans les Isles Britanniques, après en avoir suivi les traces presque dans toute l'Ecosse, au Nord, à l'Ouest au Centre et au Sud de l'Irlande et dans tout le Nord de l'Angleterre. J'ai retrouvé les mêmes surfaces polies qu'en Suisse, les mêmes moraines latérales et terminales, la même disposition *rayonnante* du centre des chaînes de montagnes vers la plaine, les lacs partout également protégés contre le remplissage par les glaciers qui en occupaient le fond.

Je me suis assuré que toutes les routes paralléles de Glen Roy et de Glen Spear ont été produits par des lacs flottant des glaces et barrés par de grands glaciers dont on voit encore la direction aux traces qu'ils ont laissés au fond des vallées, comme si les glaciers d'Argentière et des Bossons barraient la vallée de Chamounix au-dessus et au-dessous du Prieuré de manière à transformer la vallée en un lac. Le fond de Glen Spean est strié *transversalement*. Le fait le plus extraordinaire, l'absence des deux routes paralléles supérieures dans la partie orientale de Glen Spean se trouve maintenant expliquée !

J'ai accumulé tant de preuves que personne en Angleterre ne

doute maintenant que les glaciers n'y aient existé, et ceux là qui en ont le plus vu ont été convaincu les premiers: Sabine, Sir George Mackensie. Je n'ai trouvé d'opposition que contre l'extension que je leur attribue, encore cette opposition ne s'appuie-t-elle déjà plus que sur l'invraisemblance, quelques uns disent l'impossibilité physique d'un refroidissement temporaire assez considérable pour avoir couvert l'Europe d'une calotte de glace. Cependant j'ai observé mes surfaces polies et stries *jusqu'au niveau de la mer* sur toute la plaine qui s'abaisse d'Enniskillen vers Dublin; là les stries sont dirigées du N. O au S. E., puis sur la côte occidentale d'Ecosse où je les ai même vu plonger sous la mer, elles vont du N. E. au S. O. dans certaines vallées et du S. E. au N. O. dans d'autres; sur la côte orientale d'Ecosse elles vont de l'Ouest à l'Est le plus souvent. Dans l'intérieur j'en ai vu qui étaient dirigées du Nord au Sud, et ailleurs d'autres marchant du Sud au Nord. Notez bien que partout la direction des stries et des moraines indique une marche centrifuge, et nulle part un refoulement allant des côtes de la mer à l'intérieur des terres. Impossible dès lors de songer à des courants. Si l'on pouvait penser à un réhaussement du sol, les lacs et les routes paralléles s'y opposeraient, et pour cela d'ailleurs il faudrait un soulèvement simultané des montagnes partout où le phénomène a été observé, ce que la géologie dément.

Les observations paléontologiques de Mr. James Smith de Lardenhill ne contribueront pas peu à établir ma théorie. Il vient de découvrir une faune *arctique*, sur les bords de la Clyde, dans les limons superposés aux détritus des glaciers, à 40, 50, 80 pieds au-dessus du niveau de la mer. Les espèces sont identiques avec celles qui vivent maintenant au détroit de Behring, et différent complètement de celles qui vivent sur les côtes d'Ecosse.

Les observations d'Herschel sur les étoiles variables et périodiques pouront peut-être rendre un jour compte de ce refroidissement.

Je suis désolé d'être obligé de m'occuper maintenant de Poissons fossiles et de devoir laisser vieillir toutes les observations que j'ai faites sur ce sujet, tant pendant ma course dans les Alpes au mois d'Août que dans mon voyage en Angleterre, mais je ne céderai pas à la tentation et les "Poissons fossiles" s'achèveront avant que je retourne aux glaciers, sauf une apparition que je compte y faire au plus fort de l'hiver pour vérifier quelques signaux. Un heureux événement m'a un peu remonté du découragement sous l'influence duquel je vous écrivis de Glasgow. J'ai vendu les dessins originaux de mes Poissons Fossiles,[1] en sorte que j'aurai quelques mois exempts d'inquiétudes.

J'espère que vous avez reçu mes "Etudes sur les Glaciers" ; ne les jugez pas trop sévèrement comme livre; je suis trop peu au courant de ce qui s'est fait en physique pour avoir pu tenir compte de tout ce que l'on sait et éviter les redites; mais du moins j'ai observé avec tout le soin dont j'étais capable et j'ai la conscience d'avoir éloignée toute idée systématique dans l'exposition des faits pour être plus libre de me donner carrière dans le dernier chapitre. Vous me rendriez un grand service en m'écrivant bien franchement ce que vous en pensez quant au fond; j'ai pris l'habitude de profiter des critiques et quand elles viennent d'un ami comme vous, ce sont de véritables bienfaits.

Je vous adresserai par la première occasion les Comptes Rendus des séances de la Société Géologique de Londres, que Buckland m'a remis pour vous et où vous trouverez quelques autres détailes sur la question des glaciers.

Il parait qu'Élie de Beaumont veut s'obstiner à nier même les faits les plus évidents. C'est ainsi qu'il m'affirmait l'autre jour à Paris que les roches polies et striées qui se trouvent *sous les glaciers mêmes* et dont la direction coincide avec le mouvement actuel des glaciers avaient déjà la même apparence *avant* la formation des glaciers. Des masses d'un pareil poids ont donc pu se mouvoir pendant des milliers d'années sur un calcaire aussi mou que celui de la vallée de Rosenlaui sans déranger un atôme de matière!! Puis c'est le courant de l'Ober Hassli qui en bondissant de Meyringen a creusé le lac de Brienz et d'un second coup celui de Thun!! Où donc naissaient tous ces courants alpins pour se verser à la fois au Sud, à l'Est et au Nord avec une vélocité suffisante pour lancer sur le Jura des blocs de 60,000 pieds cubes! M. de Beaumont prétend

[1] Lord Francis Egerton, a relative of Sir Philip Egerton, made the purchase and generously presented them to the British Museum.

que ce sont des débâcles de glaciers; mais alors ce devraient être des glaciers plus considérables que maintenant et il devait y avoir des glaciers partout où le phénomène des blocs erratiques se présente avec les mêmes caractères qu'en Suisse. Au lieu de réfuter ma théorie; celle de M. de Beaumont la suppose comme antécédent, c'est-à-dire qu'elle n'embrasse qu'une petite partie du phénomène, celle du retrait successif des glaciers.

Peu s'en est fallu que Murchison ne m'ait dévancé dans la découverte des glaciers en Ecosse. Dans son système Silurien il suppose qu'il a dû exister de grandes étendues de glaces qui auraient charrié les graviers et les blocs soit-disant diluviens, mais il n'a pas songé à en chercher les traces. Et chose curieuse, durant nos discussions personne ne s'est opposé plus obstinément que lui à l'existence des glaciers, qu'il a cependant fini par admettre aussi.[1]

Au moment où j'ai quitté Londres, Buckland partait pour le pays de Galles où je n'ai pu aller et où il trouvera certainement des choses curieuses. Mais j'oublie que l'hiver approche et que déjà vous devez avoir à Berlin plus de glaces que vous n'en voulez sans celles dont je viens de vous charger à profusion. Je n'ose rien vous dire pour M. de Buch quoique je l'aime toujours de tout mon cœur, on m'a dit qu'à Erlangen (Société Allemande des naturalistes) il s'était faché tout rouge contre moi parce que je fais les glaciers assez grands pour fournir de l'eau nécessaire à ses courants.

Adieu, mon bien cher ami, écrivez moi bientôt quelques lignes, vos lettres sont toujours pour moi des trésors, car elles me donnent

[1] It seems that Murchison, a short time afterward, again changed his views, and returned to the floating iceberg and mud current theory; for in his "Geology of Russia," 1845, he rejected "the glacier theory," explaining the Scandinavian drift and erratic blocks in Russia by *traînées* under the sea, made by "moistened masses of drift, under powerful causes of translation"; and in his address at the anniversary meeting of the Geological Society of London, 1842, he says: "The existence of glaciers in Scotland and England is not, at all events, established to the satisfaction of what I believe to be by far the greater number of British geologists." It was not until more than twenty years after Agassiz's visit of 1840, that at last, in 1862, Murchison wrote him that he was wrong in opposing as he did the glacial period. He took time to consider!

ce courage et ce contentement d'esprit sans lesquels on ne fait rien de bon. Ce qui me fait surtout croire que j'ai bien vu en Ecosse, c'est que c'est à vous que je rendais compte mentalement de mes observations.

Votre tout dévoué pour la vie,

LOUIS AGASSIZ.

CHAPTER VIII.

1841–1842.

VISIT DURING THE WINTER TO THE AAR GLACIER — LETTERS to JULES THURMANN AND TO EUGENIO SISMONDA — "MONOGRAPHIE D'ECHINODERMES VIVANTS ET FOSSILES" — LETTER TO DESHAYES — ANOTHER LETTER TO THURMANN — VISIT OF JAMES D. FORBES AT THE "HÔTEL DES NEUCHÂTELOIS" — ASCENT OF THE JUNGFRAU — OTHER VISITORS AT THE "HÔTEL DES NEUCHÂTELOIS" — FORBES AT NEUCHÂTEL AND LA CHAUX-DE-FONDS — INAUGURATION OF THE ACADEMY OF NEUCHÂTEL, 18TH OF NOVEMBER, 1841 — AGASSIZ'S LETTER TO THE RECTOR OF THE ACADEMY — HIS APPOINTMENT AS RECTOR FOR THE YEAR 1842–1843 — CONTROVERSY WITH JAMES D. FORBES ON THE LAMINATED STRUCTURE OF GLACIERS — A NEW CABIN TO REPLACE THE "HÔTEL DES NEUCHÂTELOIS" — STAY AT THE AAR GLACIER FROM THE BEGINNING OF JULY, 1842, TO THE MIDDLE OF SEPTEMBER — DISCOVERIES OF JOHN TYNDALL — DISPUTE WITH KARL SCHIMPER — DANIEL DOLLFUS-AUSSET.

THE winter of 1841 was so rainy at Neuchâtel, and in consequence so much fresh snow fell on the Oberland Alps, that Agassiz was obliged to postpone his proposed visit to the "Hôtel des Neuchâtelois" until the 8th of March. On that day he left Neuchâtel with his secretary, reaching the Grimsel three days later, without very great difficulty. An hour before their arrival, the guardian of the hospice was advised by the movements of his dog, a fine and very large Newfoundland, that some one was approaching. As is often the case in the Alps and mountainous country, the temperature was higher at the Grimsel than at Interlaken. The amount of snow was enormous; the hospice was buried in it; and when the travellers, after a rather exhausting walk, reached the place where the "Hôtel des Neuchâtelois" should have been, they were greatly surprised to see nothing of it but a sort of hump on the crest of snow which covered the moraine. However, after forcing their way around this hump, they found on one side a few feet of the big boulder. It was impossible to enter it without clearing away an enormous mass of snow; so Agassiz contented himself with lying down on the snow, and enjoying the marvellous spectacle around him. The weather was perfect; the air so clear that every topographical feature of the Finsteraarhorn and other peaks was seen with a distinctness unknown during the summer season. The travellers went as far as the Abschwung, then returned to the place of the "Hôtel des Neuchâtelois," where they saw the tops of two very high stakes placed there in the preceding August in holes bored into the ice. Agassiz remained behind with one guide to make several observations with a thermometrograph, and finally returned to the Grimsel, after a journey of twelve hours, from 4 o'clock A.M. to 4 o'clock P.M., somewhat tired, but very happy in his success; for he was certainly the first visitor to the Aar Glacier in the winter season. From the Grimsel Agassiz crossed by Meyringen to Rosenlaui, where he visited the glacier to examine the polishing of the rocks in contact with the ice, and also to determine the quantity of water arising from the glacier. And in regard to the latter point, like de Saussure at the glacier des Bois at Chamounix, he concluded that during the winter the glacier yielded only spring water. A week after leaving Neuchâtel they returned home rather sunburned by their exposure to the intense sunlight on the snow-field they had travelled over.

We have seen in the last letter to Humboldt that Agassiz gathered a large collection of fossil echinoderms during his stay in England in 1840. He had done the same in passing through Paris, and was very diligent in getting specimens from every geologist living among the Jura Mountains,—as Thurmann of Porrentruy, d'Udressier and Parandier of Besançon, and Merian of Bâle. The following letter to Jules Thurmann gives some rather curious details:—

LE 7, 1840 or 1841 ? (date not distinct).

Monsieur,—Voici la première livraison de mes Echinodermes, j'y joins la première des *Études critiques sur les Mollusques*, quoique le texte ne soit pas encore prêt, dans l'espoir que vous aurez peut-être à me communiquer quelques Trigonies ou Myes que je n'ai pas et que je pourrais encore ajouter à mes planches. Le prix des Echinodermes est de 10 francs, je réclamerai celui des Mollusques en vous envoyant le texte.

Je viens de faire demander à M. Nicolet les deux premières livraisons de Sowerby, la 3ième est très avancée; le prix de la livraison coloriée est de 10 francs. M. Nicolet vous enverra lui-même par occasions les livraisons.

Je pense que vous apprendrez avec plaisir que Gressly a repris son activité d'autrefois; j'ai reçu depuis peu plusieurs bonnes lettres de lui. Excusez-moi de tant tarder de vous envoyer mes moules; j'ai encore eu des chagrins de famille cet hiver qui m'ont fait passer plusieurs semaines en Allemagne et singulièrement dérangé mes affaires; dès que je le pourrai je réparerai mes torts.

Votre très dévoué serviteur,

AGASSIZ.

The following is another letter written about the same time: —

MONSIEUR EUGÈNE SISMONDA,
assistant au Musée royal de Minéralogie, Turin.

Monsieur et très honoré collègue, — De retour d'Angleterre après une absence de près de quatre mois j'ai le plaisir de recevoir votre aimable lettre. Je suis charmé d'entrer en relation directe avec vous, qui par vos beaux travaux géologiques avez si puissament contribué à l'avancement de la science. Il y aura tout profit pour moi à soutenir une correspondance suivie avec vous. J'accepte avec plaisir votre proposition d'échange; je puis vous remettre au moins 600 moules d'Echinides fossiles accompagnés d'un catalogue systématique et d'une caractèristique des genres nouveaux que j'ai établis. Je recevrai volontiers en échange des coquilles, des Zoophites et des Echinides de tous vos terrains d'Italie, même les espèces les plus communes. Je désire beaucoup obtenir des séries d'exemplaires de différents ages.

Dans des échanges de ce genre j'ai généralement demandé un fossile contre un moule à raison des frais considérables que leur exécution m'a occasionné, sans compter jamais rigoureusement, comme cela convient entre gens qui doivent avoir en vue les intérêts de la science plutôt que la dépense qui en résulte pour eux. Outre ces moules d'oursins j'en possède beaucoup de Mollusques, de Poissons, de Mamifères et 5 à 600 espèces de Mollusques des terrains secondaires en nature, dont je puis disposer pour échanges. C'est assez vous dire que j'ai d'amples matériaux pour des envois considérables et j'attends seulement pour vous expédier une première caisse qui peut être préte dans trois jours, que vous vouliez bien me dire quel nombre d'exemplaires vous avez de disponible, ou plûtot quelle étendue vous désirez que je donne à mon premier envoi.

Je suis très flatté de la dédicace que vous me faites d'un de vos Echinides et je me réjouis à l'avance d'apprendre à le connaître.

Agréez, Monsieur, l'assurance de ma considération, très distinguée.

LS. AGASSIZ.

NEUCHÂTEL EN SUISSE,
le 24 décembre, 1840.

We have in these two letters a glimpse of Agassiz's method of collecting specimens, making exchanges, and disposing of his publications.

The success of the "Fossiles du terrain crétacé du Jura Neuchâtelois," of the "Prodrome d'une Monographie des Echinodermes," and of "Echinodermes fossiles de la Suisse," all published at the expense of the Société des Sciences Naturelles de Neuchâtel and of the Société Helvétique des Sciences Naturelles, led Agassiz to undertake, at his own expense, the publication of "Monographies d'Échinodermes vivants et fossiles," with many beautifully executed plates. It was an unfortunate undertaking, very expensive on account of the great number of plates, and without proper patronage from naturalists to make it profitable. Only four monographs or "livraisons" were issued between 1838 and 1842. The first, on "Salénies," 1838, shows good work, and is very creditable in all respects, and worthy of the name which signs it; the second, on "Scutelles" (July, 1841), although containing many new facts and an interesting history of the progress of the natural history of the echinoderms, besides twenty-seven most exact and beautiful plates, did not attract much attention; while the third "livraison," containing the "Galerites" and "Dysaster" (1842), is by E. Desor. Agassiz helped in the revision of the proof-sheets; but, on the whole, the work shows a noticeable inferiority to all the previous publications on the echinoderms.

The fourth "livraison" (1842), the manuscript of which was written in German and translated into

French, treats the anatomy of the genus Echinus, and is by Professor G. Valentin of Berne; and its nine plates, several of them double, are remarkably well drawn by Dickmann. After the issue of the fourth "livraison," the publication was stopped and never resumed. This fine work, forming a large 4to volume, is dedicated to "M. Valenciennes, Professeur de Zoologie au Jardin des Plantes et à M. Paul Deshayes, Professeur de Conchyliologie à Paris." In this way Agassiz tried to conciliate two naturalists, who had nothing in common except a disagreement in regard to an appointment obtained by pure favour for Valenciennes, against all justice and the right claim of Deshayes. For, through the influence of Humboldt and the help of Agassiz, Valenciennes was elected professor of conchology and zoöphytology at the Jardin des Plantes, — a most unfortunate choice, for he knew next to nothing of these two difficult branches of invertebrate zoölogy, having only a knowledge of living fishes, obtained as an assistant of George Cuvier; while Deshayes, on the other hand, was regarded by every naturalist, not only in France, but also in other countries, as the ablest conchologist of his time.[1] Agassiz, hoping to mend matters and to help in healing the wound inflicted on Deshayes, conceived the strange notion of uniting in a dedication the two names of Valenciennes and Deshayes, placing Valenciennes before Deshayes. He very well knew that he was

[1] Thirty years later, in 1869, Deshayes was at last appointed Professor of Conchology at the Jardin des Plantes, at the ripe old age of seventy-two years, an act of justice due to M. V. Duruy, then Secretary of Public Instruction.

treading on dangerous ground, as the following letter to M. Deshayes shows: —

NEUCHÂTEL, le 27 février, 1839.

Monsieur, — Désirant vous donner un témoignage public de ma reconnaissance pour les communications importantes que vous m'avez faîtes sur les oursins fossiles, j'ai pris la liberté de vous dédier conjointment à M. Valenciennes à qui je dois également des communications d'une haute valeur, l'ouvrage sur cette classe d'animaux, dont je viens de publier la première livraison.

J'espère, Monsieur, que vous daignerez accepter cette marque de mon estime et de mon amitié. La science vous doit de si importants travaux, trop peu récompensés, dans l'atmosphère où vous vivez, pour que je ne puisse pas espèrer trouver de la sympathie chez un homme qui poursuit ces recherches avec un tel désinteressement. La deuxième livraison, qui est très avancée, contiendra les Scutelles. Je profiterai de toutes les occasions que j'aurai pour Paris pour vous retourner vos exemplaires au fur et à mesure qu'ils seront dessinés, j'y joindrai les moules des espèces que vous n'avez pas et si vous le désirez de celles des vôtres que j'ai pu faire mouler sans risque de les endommager. Ils pourraient vous servir à faire des échanges. Veuillez me dire si vous désirez que je vous en fasse couler des épreuves.

Je vous adresse également la première livraison encore inachevée d'un ouvrage que je prépare depuis longtemps sur les Mollusques fossiles de la Suisse principalement, dans lequel je me propose de traiter aussi différentes questions générales de Conchyliologie et surtout celle de la délimitation des genres et de l'analogie des espèces fossiles avec les espèces vivants. Quoiqu'envisageant, comme vous le savez ces questions un peu différemment de vous, la base sur laquelle j'ai travaillé n'en est pas moins la même et c'est là un point de ralliement infaillible l'étude consciencieuse et comparative des faits. Qu'après cela il me paraisse plus utile de grouper les espèces d'après leurs caractères plus restreints en petites groupes, que de les réunir d'après des caractères plus généraux en grands genres, c'est une question à débattre ultérieurement et le résultat auquel on s'arrêtera ne changera en rien la valeur des observations spéciales.

La définition et la circonscription des espèces touche déjà de plus près à l'importance actuelle que l'on attache à ce genre de travail. Il me paraît à cet égard que la facilité de distinguer telle ou telle série de formes diverses ne peut pas être un motif absolu pour les réunir ou les séparer et qu'il importe de rassembler les matériaux les plus complets sur la généalogie de chaque type avant de pouvoir se prononcer d'une manière invariable. C'est ainsi que la possibilité de rattacher à une souche primitive les générations actuelles souvent diverses de telle ou telle espèce fait que nous ne les séparons pas comme autant d'espèces distinctes, bien que souvent les individus que nous réunissons ainsi différent d'avantage entre eux que ceux d'autres types que nous séparons à cause de la fixité de leurs caractères. Ces principes de la zoologie actuelle me paraissent devoir influer ultérieurement sur notre manière d'envisager l'analogie des espèces fossiles avec les vivantes. Je crois par exemple que s'il pouvait être démontré géologiquement que certaines espèces fossiles que nous envisageons comme identiques avec les vivantes, ont cessé d'exister dans des circonstances telles qu'il serait impossible qu'elles aient pu se reproduire par voie de génération dans l'époque suivante, il faudrait alors envisager ces analogues d'une autre époque comme des espèces particulières procréez dans d'autres temps alors même que leur ressemblance extérieure rendrait leur distinction très difficile. Il me semble qu'en pareil cas le fait que les extrêmes des variétés d'un type fossile se lient aussi étroitement aux extrêmes d'un type vivant que leurs variétés entr'elles n'emporte pas la nécessité de la réunion des deux types en une seule espèce. Quoique cette manière de voir ne s'accorde pas en tous points avec certains principes que vous avez établis sur la détermination des espèces, ils ne me paraissent infirmer en aucune façon l'importance des faits innombrables que vous avez receuillis sur l'analogie des espèces fossiles avec les vivantes, puisque vous avez toujours signalé les particularités qui distinguent toutes les variétés que vous avez réunies dans le même type spécifique. Je pense dès lors que vous ne répugnerez pas à faire part vous même occasionellement de ces observations à la Société Géologique (de France).

La première livraison de mes "Études critiques" paraîtra dans le courant de l'été. En parcourant ce que j'ai pu vous en envoyer dès aujourd'hui, vous remarquerez sans doute que j'ai pris toutes les précautions possibles pour éviter de multiplier les espèces sans raison; ainsi pour les Trigonies j'ai représenté une série de tous les âges de la *Trigonia navis* (du Lias Supérieur de Gundershofen, Haut Rhin, recueillis par Gressly) pour prouver que quelques espèces nouvelles que j'ai établies n'en sont pas les jeunes. Quant à la famille des Myes elle a eu des représentants bien plus nombreux et plus variés dans les terrains jurassiques, qu'à des époques plus récentes, et la diversité des types que j'ai étudiés m'a engagé à grouper des espèces (qui cadraient fort mal dans les genres admis maintenant) dans plusieurs genres nouveaux dont je donnerai très en détail les caractères distinctifs en les comparant soit entr'eux, soit avec des genres qui ont des représentants maintenant.

Si votre Conchyliologie continue à paraitre régulièrement j'attendrai que vous ayez publié les Myes pour émettre ma Monographie. Je suis enchanté des deux livraisons qui ont déjà paru.

Je vous adresse enfin les premières planches d'un Mémoire[1] que je vais insérer pour paraître incessamment dans le second volume des Mémoires de la Société d'Histoire Naturelle de Neuchâtel. Je pense que cette publication sera utile pour la détermination des moules fossiles. Si vous désirez les avoir en plâtre, je vous les enverrai, mais comme ils appartiennent à notre Musée, vous m'obligerez de m'envoyer en échange des fossiles ou des coquilles vivantes. Dans ce cas je vous indiquerai ce qui nous manque surtout.

Agréez, Monsieur, l'assurance de mon parfait dévouement,

LS. AGASSIZ.

This is one of the most important scientific letters written by Agassiz, showing the direction of his mind and his preconceived ideas on a subject which he advocated, more or less, during his whole life, in regard to species and genera, and also the erroneous notion of the confinement of species to each group of formations,

[1] "Mémoire sur les Moules de Mollusques vivans et fossiles, première partie; Moule d'Acéphales vivants." 4to, 1839.

against the plain facts brought forward by Deshayes, demonstrating the passage of species from one group to another, illustrated so vividly in the great Tertiary epoch of the Paris basin. In palæontology Agassiz was an absolutist until the last two years of his life, when he abandoned the rigidity of his principles in his celebrated prophetic letter to Benjamin Pierce on the supposed existence of Ammonites, Trilobites, and other lost forms of marine animals at great depths. These two errors are the most remarkable examples of the excess of his imagination.

Another scientific letter, written at the same time to Jules Thurmann, is too good not to be given in full.

NEUCHÂTEL, 12 février, 1842.

MONSIEUR JULES THURMANN
à Porrentruy.

Monsieur,—Gressly m'a fait le plaisir de me communiquer la lettre que vous lui avez adressée tout récemment. Je me réjouis infiniment d'apprendre que vous vous êtes remis avec ardeur à la Géologie, et que vous étudiez maintenant sérieusement les fossiles. Je vous remercie infiniment pour ma part des détailes circonstanciés dans lesquels vous êtes entré sur les oursins, et rien ne me serait plus utile et agréable que de recevoir vos observations sur les autres parties de mon travail. Soyez persuadé, Monsieur, que j'apprends bien davantage des remarques de ce genre, que les compliments d'une banalité affligeante, que les auteurs s'adressent si souvent. Je ne puis même vous prouver l'importance que j'y attache, qu'en répondant à vos remarques.

Une chose m'a frappé, c'est que mes coupes génériques vous aient satisfaits. À ce sujet, je suis de la part des Zoologistes en bute à des reproches continuels; on me répéte sans cesse que je les multiplie à plaisir. Quant à moi j'ai la conviction que l'on ne parvient bien à étudier les espèces, qu'en les groupant dans des genres aussi restreint que possible; sauf, peut-être à en réunir plus tard plusieurs sous un même chef, si l'on découvre des types intermédiaires.

Quant aux espèces je partage pleinement les *principes* que vous énoncez, je les professe hautement; je dirai même que ce sont ces principes qui me dirigent dans mes études; mais je diffère surtout de vous dans leur *application*. Ayant égard à ce qui a eu lieu dans les autres branches de la science, lorsqu'elles étaient dans leur enfance, je cherche à réunir le plus de matériaux possible, et après avoir comparé exactement, je distingue et distingue, faisant valoir les moindre différences que j'aperçois, établissant des espèces souvent d'après un seul exemplaire imparfait, sauf à réunir, quand on a rassemblé des matériaux suffisants pour le faire à bon escient; c'est la marche que la science a suivie dans toute son histoire. Ce travail de critique a ses inconvénients, je le sais; il oblige de revenir sur les mêmes matériaux à plusieurs reprises; mais il a ses grands avantages, c'est de forcer à un examen scrupuleux, tous ceux qui réunissent des matériaux nombreux sur une seule espèce. Il forcera les collecteurs à ne pas disséminer à l'infini leurs exemplaires et à collecter des séries et non pas des échantillons. Vous verrez que toutes les fois qu'il m'a été possible d'étudier des séries d'exemplaires, j'en ai analysé toutes les formes, distingué des variétés d'âges, de station, etc. Vous reconnoissez vous-même, que nos collections sont trop pauvres pour nous permettre de faire cela, pour un grand nombre d'espèces dès à présent. Le terme désirable n'est donc pas encore à notre portée, et voilà pourquoi je procéde si différemment de la plupart de mes collègues dans l'application des principes incontestables. Pour les mêmes raisons j'ai fréquemment établi des genres dont je n'ai longtemps connu qu'une seule espèce.

Maintenant vous possédez des séries d'espèces dont je n'ai vu jusqu'ici que des exemplaires isolés; c'est une bonne trouvaille et si vous voulez bien me les communiquer au complet, c'est-à-dire la masse des bons exemplaires je serai le premier à supprimer celles de mes espèces qui forment double emploi. Mais avouez, qu'il était plus profitable à la science, que c'était du moins fixer les yeux ou l'attention d'une manière bien plus pressante sur ces oursins, en établissant le genre *Pedina* et en y distinguant plusieurs espèces qu'en les réunissant sous un seul nom dans le genre *Echinus*. Supposons un instant, qu'au lieu de devoir les réunir, en partie du moins, comme cela me paraît probable, d'après ce que vous écrivez à Gressly, les ayant d'abord réunis, quelqu'un eut trouvé que l'on confondait plusieurs espèces sous le même nom. La vraie difficulté qui se serait alors présentée eût été de savoir à laquelle, il faut conserver le nom primitif; puis si cette espèce est établie depuis longtemps, s'assurer laquelle des deux est mentionnée dans les différents auteurs; puis effacer toutes les citations de localités déjà mentionnées, parce qu'on ne sait plus, quelle est celle qui provient de l'endroit A. ou de l'endroit B., etc. C'est-à-dire que c'est à *la crainte d'établir trop d'espèces*, sur des matériaux incomplets, qu'il faut attribuer tout cet effroyable dédale de la synonymie, et des fausses citations de gisements, qu'on ne peut éviter qu'en s'abstenant complètement, ce qui ne fait faire aucun progrès, ou en distinguant et distinguant toujours jusqu'à ce qu'on puisse réunir à coup sûr.

N'en n'a t'il pas été ainsi de tous nos oiseaux aquatiques, dont les jeunes et les vieux ont passés pour des espèces distinctes, même aux yeux de Linnée? C'est cette manière d'agir qui m'a conduit à établir bien des espèces, qu'il faudra peut-être supprimer un jour. J'ajouterai encore que c'est faute de posséder moi-même une grande partie des objets que je décris, que je suis forcé de faire faire les planches, pendant que ces objets sont à ma disposition et souvent de les numéroter d'A. B. C. et de faire pire encore. Aussi si nous nous voyons plus souvent, m'entendriez vous souvent répéter qu'il ne faudrait jamais publier que la seconde édition de ses œuvres et toujours canceller la première après en avoir fait part à ses amis seulement. Nous ne marcherons avec une entière assurance en paléontologie, que quand on possédera autant d'éditions d'un genre complet du Règne Animal, qu'il y a d'éditions de "Cornelius Nepos" ou de la grammaire latine de Bröder, ou de tel dictionnaire de poche.

Gressly, Desor et moi nous travaillons aussi assiduement que possible à la paléontologie; ces Messieurs vous écrivent chacun de leur côté; j'ai voulu aussi vous donner un signe de vie, et j'espère que nous n'en resterons pas là. Jent [l'éditeur d'Agassiz à Soleure] ne peut pas vous avoir adressé les planches de Myes, puisqu'il ne les a pas encore. Ce sera Gressly ou moi qui vous les auront fait parvenir, afin d'apprendre de vous, si vous aviez quelque chose de neuf dans ces genres. Vous m'obligerez infiniment en en faisant une petite caisse et en me l'adressant avec vos oursins. Nous déterminerons cela en commun, et Gressly ou Desor vous renverra prochainement le tout.

Agréez, Monsieur, mes salutations très empressées.

LS. AGASSIZ.

Seldom has Agassiz written such an important scientific letter, showing as it does his method of determining species and establishing new genera. In it he anticipates the criticism which has been made since, that he created too many species; for instance, of fossil fishes from the *flysch*[1] of Glaris, some of which were distorted by strong lateral pressure. A larger number of specimens than were then at his disposal have since proved his mistake.

At the meeting of the British Association at Glasgow during September, 1840, James D. Forbes, professor of natural philosophy in the University of Edinburgh, took such an interest in all the communications made by Agassiz on the glacial question and the glaciers, that Agassiz very politely tendered him an invitation to visit him the next summer at his "Hôtel des Neuchâtelois."

On the 8th of August, 1841, Agassiz, with his assistants, again occupied their old and rather rough quarters on the Aar glacier; and there Forbes, accompanied by a Scotch friend, Mr. Heath, was received as a welcome guest. Agassiz was delighted to have a physician so celebrated as Forbes to examine his observations. He showed him everything — all the experiments they were

[1] A lithologic term used in German Switzerland to designate a series of strata belonging to the Tertiary Eocene.

making in regard to temperature, progress of the moraine, etc., often asking his opinion and advice. But Forbes was as silent as a sphinx; it was impossible to draw from him a single remark or hint. This impenetrability in a savant was new to Agassiz, who, until then, had more or less easily charmed every scientific man with whom he had come in contact. But this time he had found one who would not yield to his ingenuousness. During the three weeks spent by Forbes at the "Hôtel des Neuchâtelois," he observed everything around him, but said absolutely nothing, even as regards his impressions. Agassiz's desire to study the structure of glaciers led him to bore in the glacier a hole 140 feet deep; and he was also lowered, supported simply by a rope, to a depth of 120 feet, into an old "moulin" or well, to see how far through the glaciers the laminated structure extended. This veined structure was the only point referred to by Forbes during his stay at the "Hôtel des Neuchâtelois." It had been observed previously by David Brewster, Hugi, Bishop Rendu, Guyot, and Agassiz; but Forbes afterward claimed that it was he who first called Agassiz's "attention to the fact that the ice of glaciers is composed of vertical laminæ, constituting a true ribboned structure,"[1] and raised a controversy, of which we shall speak farther on.

Several peaks were ascended by Agassiz during Forbes's stay, among them the summit of the Ewigschneehorn; with a visit to the Gauli glacier, a walk over the "mer de glace de Viesch," and, finally, an

[1] "Edinburgh Philosophical Journal," January, 1842.

ascent of the Jungfrau. Until then no tourist had succeeded in reaching the top of the Jungfrau. During the last two years Agassiz had often discussed with his favourite guide, Jacob Leuthold, the means of reaching that virgin peak, the great landmark of the Bernese Oberland.[1] On the 27th of August, Agassiz with Forbes, Heath, Desor, and two others, and six guides, left the Grimsel at four o'clock in the morning, arriving at six o'clock P.M. at the Meril Chalets, near the lake, where they were well received by the herders, who were rather astonished at the arrival of such a large party.

Next morning, at five o'clock, they left Meril, aiming for the Aletsch glacier; after a fatiguing walk on rather slippery ground, among "crevasses" and over snow fields, the party reached the base of the last slope at three o'clock P.M. Four of the party had been forced by fatigue or giddiness to remain behind; but the other eight — one after another in turn — gained the summit, which is only two feet long by a foot and a half broad. Agassiz was the first, then Desor, Forbes behind, and a French tourist, M. Duchâtelier of Nantes, fourth. At four o'clock the descent began; and they arrived all safe at half-past eleven P.M. at the Meril Chalets. Three days later Agassiz was again at the "Hôtel des Neuchâtelois," where he found his artist-friend, Burkhardt, and his assistant, Charles Girard, anxiously awaiting his return.

[1] The two brothers, Rudolph and Jerome Meyer of Aarau, in 1811 and 1812, made two ascents of the Jungfrau with success, although the fact is contested by the mountaineers of the country. At all events, a party of guides, with J. Baumann as chief, succeeded in reaching the summit on the 8th of September, 1828.

During the five weeks' sojourn of Agassiz and his friends on the glacier of the Aar, from the 8th of August to the 10th of September, 1841, many visitors were received besides Forbes. The title of "Hôtel des Neuchâtelois" deceived several tourists, who, hearing of it at the Grimsel Hospice, came up expecting to find some establishment like the "Culm Hotel" on the Rigi. Even a Scotch lady, Mrs. Covan of Edinburgh, in returning from an ascent of the Finsteraarhorn, stopped and was entertained as well as it was possible by Agassiz. Most of these visitors were obliged to return to the Grimsel to find lodging, or to be contented with a corner in the guide's cabin. The hospitality of the "Hôtel des Neuchâtelois" was reserved for savants or personal friends, such as General de Pfuel, the Prussian governor of Neuchâtel, Lord Enniskillen of Ireland, the two de Rougemont of Neuchâtel, the geologists, Studer and Escher von der Linth, the meteorologist and botanist, Charles Martins, Bravais, Guyot, etc. However, the solidity of the block forming the roof was beginning to awaken suspicion; cracks had become alarmingly numerous, and when it rained, the interior of the hotel was almost a pond, with water running in every direction. It was only a question of time when the enormous block would break in pieces, and it was also feared that a sudden move of the glaciers might hasten the catastrophe. So every evening before retiring one of the party used to make the round of the cabin to see that all was right. Although not in immediate danger, it was resolved to abandon the "Hôtel des Neuchâtelois" and to erect next year a new cabin; not, as formerly, on the glacier, but on firm ground, and hence less exposed to dangerous accidents.

Forbes, after his return from the ascent of the Jungfrau, visited parts of the Valais and Chamounix, and by the middle of September arrived at Neuchâtel. His reception by Agassiz was most cordial; and Agassiz's letter introducing Forbes to his good friend, Célestin Nicolet, may be quoted as an evidence of his solicitude to help him in every way.

NEUCHÂTEL, le 20 septembre, 1841.

Mon cher ami Nicolet.—C'est M. le Professeur Forbes qui vous remettra ces lignes et que je vous recommande tout particulièrement en vous priant de lui faire voir ce que vous avez d'intéressant à la Chaux-de-Fonds, en fait de Sciences et d'Industrie. Je suis sûre que vous aurez beaucoup de plaisir à faire la connoissance d'un homme aussi haut placé dans le monde savant que M. Forbes.

Venez bientôt causer un peu plus intimément avec nous de tout ce que nous avons vu dans les Alpes; plus vite et mieux.

Tout à vous

LS. AGASSIZ.

Although the Neuchâtel Academy was founded in 1838, the public inauguration of the new institution was delayed until the 18th of November, 1841. Agassiz on that occasion, wearing the cross of the Red Eagle of Prussia, delivered an address, "De la Succession et du développement des êtres organisés à la surface du globe terrestre dans les différents âges de la nature," in which he says, "Si le cours des astres ne nous présente aucune variation, si l'ordre des saisons est immuable, si la reproduction des espèces s'opère toujours de la même manière, il est évident que le cours de ces phénomènes est invariablement réglé et suit des lois naturelles, indépendantes de l'influence créatrice qui les a établies."

Objections were raised by the rector of the academy and some of the professors; and after discussion, it was resolved that two hundred separate copies of Agassiz's address should be printed for his own use, and four hundred copies of the three speeches delivered at the inauguration, as part of the programme and annual report of the academy. The pietist party was very strong then in Neuchâtel, and several sentences in Agassiz's address were considered as anti-orthodox and antagonistic to the prevailing creed of the Neuchâtel ministers. The following letter from Agassiz to the rector shows the intensity of the commotion produced: —

NEUCHÂTEL, 14 décembre, 1841.

Mon cher collègue [le recteur Pétavel]. — Considérez, je vous prie, que mon discours s'adresse au public de l'Académie et que peu m'importe le jugement de ceux qui sont incompétents ou incapables. Que serait notre Académie si elle devait se mettre à la hauteur de tous ceux qui en veulent? Vous auriez pu voir vendredi que je fais de vos réclamations une affaire de principes et que je suis parfaitement décidé à ne pas faire la moindre concession, parce que j'y verrais une atteinte fatale à la liberté d'enseignement et parce que je tiens à ce que notre Académie aie de la tenue et qu'un de ses membres ne dise pas blanc aujourd'hui et noir demain. Ne confondez pas votre position avec la mienne; vous deviez parler au nom du corps académique et c'est ce qui nous donnait à tous le droit d'exiger que vous parleriez dans tel ou tel sens, dans celui de la majorité, sauf à vous de donner votre démission comme Recteur, s'il ne vous convenait pas d'être l'organe de notre pensée: Vous ne faites pas assez cette distinction. J'ai parlé pour moi et dans l'intérêt de notre Académie; je ne souffrirais pas la moindre critique de ce que j'ai fait et dit; je vous le répéte sans la moindre animosité; je dirai même que je le fais comme si j'étais tout à fait étranger à la discussion et uniquement parce que vous me demandez comme Recteur: quel bien je pense que cela fera à l'Académie, et parce que me faisant le défenseur de cette indépendance de l'esprit, sans laquelle rien ne grand ne peut prospérer, je dois vous rappeler que le Recteur est tenu de rester étranger à tout cela. S'il en était autrement, ce seraient des antécédents qui donneraient accès au cœur même de l'Académie à des influences étrangères, que je ferai toujours tous mes efforts d'en bannir et auxquelles il faut fermer la bouche dès le commencement pour qu'elles ne réitèrent pas leurs tentatives. Je vous l'ai dit dans mon discours et je vous le répéte ici: il est peu de grandes vérités qui n'aient été traitées de chimères et de blasphèmes, avant qu'elles fussent démontrées. Heureusement que les temps de Galilée n'existent plus; mais aussi y a-t-il bien moins de mérite qu'alors à ne pas composer avec les prétentions des *Ministres de l'Église*, et ce n'est certes pas une couronne de martyr que j'espère conquérir. Je dis "de l'Église," et par là j'entends les ministres de *tous* les cultes, qu'ils soient protestants, catholiques, juifs ou mahométans, *qui ne veulent faire de progrès en rien*. Notez bien que je ne vous dis pas "de la Religion." N'oubliez pas que mes doctrines ne peuvent porter d'atteinte qu'à l'enseignement des docteurs de l'Église et nullement aux vérités de la Religion.

J'en reviens à mon discours. Ennuyé de toutes ces discussions, je le livrerai aujourd'hui à Wolfrath (l'imprimeur) sans notes, tel que je l'ai lu, sans y changer quoique ce soît. Si on ne me laisse pas tranquille à ce sujet, ce sera ma meilleur défense.

Agréez, mon cher collègue, mes salutations bien empressées; croyez que j'estime votre zèle pour les convictions que vous professez maintenant. Soyez persuadé que jamais je ne chercherai de disçuter sur ces matières, que je désire avant tout vivre en paix avec mes convictions et pouvoir poursuivre sans relâche mes recherches, ne réclamant en leur faveur que la même tolérance que je concéde à tout le monde.

LS. AGASSIZ.

This is the most decisive letter ever written by Agassiz. At that time Neuchâtel was entirely in the hands of the "Ministres de l'Église"; and the Pietists and even the "Momiers" largely controlled Neuchâtel society. Some of Agassiz's most intimate friends, like Arnold Guyot, were among the leaders of the Pietists, and it required considerable moral courage to resist the anti-liberal pressure exerted by the sect against all liberal, even scientific ideas. This controversy is the best answer to the attacks of those who have pretended that Agassiz came under the influence of the Methodists in America. Agassiz, on the contrary, was most liberal in religion, and always took care never to confuse science with religion. All his life he kept free of the "Ministres de l'Église," both in Europe and in America. To tell the truth, he never liked "ministers," to whatever sect they might belong. To finish this incident, Agassiz was appointed rector of the Neuchâtel Academy for the year 1842–1843; and in his "Discours du Nouvel-An," the 1st of January, 1843, he said: —

Une institution aussi jeune que notre Académie a surtout besoin de l'appui d'un monarque qui veille avec une si constante sollicitude aux intérêts de la science. Déjà avant son avènement au trône Frédéric-Guillaume IV était le protecteur le plus zélé des sciences en Prusse, et sous son règne les institutions scientifiques du royaume brillent d'un nouvel éclat, dû surtout à l'empressement avec lequel le roi a appelé dans ses États les hommes les plus éminents de l'Allemagne dans le domaine de la philosophie, des sciences, du droit et des lettres. Tout récemment encore, il a inauguré l'œuvre d'une grande reconciliation religieuse. C'est lui qui a brisé de vielles entraves qui pourraient gêner un nouvel élan de l'intelligence pour agrandir les limites d'une libre expression de la pensée, tout en la contenant dans de sages bornes.[1]

[1] "La première Académie de Neuchâtel," par Alphonse Petitpierre, pp. 128–129, Neuchâtel, 1889.

In his agenda, Agassiz wrote the same day: "Présenté comme Recteur les hommages du corps académique au Roi. La réponse du Président du Conseil d'État me fait supposer que les paroles modérées que j'ai prononcées ont déplu." After a second thought, and on the advice of the governor, General de Pfuel, representing the king of Prussia, the matter was dropped.

The year 1842 began with a difficulty with James Forbes, ended with one with Karl Schimper, with the erection of a new and rather too costly establishment at the glacier of the Aar as an interlude, — three things which might have been avoided to the advantage of Agassiz. On the 26th of February his secretary, who by this time had become hardly inferior to Agassiz, wrote a rather sharp and irritating letter to Forbes, relating to the question of priority in the discovery of the laminated structure of the glacier. Desor, by inclination and education, was always ready for a controversy or a discussion on any point scientific, political, or religious. He had learned enough of law, when a student, to assimilate the spirit of the advocate. He became a naturalist by accident, and as a means of supporting himself. But his proper sphere was politics; and as soon as he became unexpectedly rich, he devoted almost all his time to politics; as his biographer says: "He was persuaded, at the end of his life, that on his shoulders rested the welfare of the Swiss Confederacy, of the Neuchâtel Canton, and of the federal Polytechnicum. He passed all his time in writing polemic articles in newspapers, compromised himself in petty personal discussions, and founded a newspaper to advocate and maintain his ideas and regain his political position, promptly lost by his own fault. In the end, his newspaper was reduced to only one hundred subscribers; and the fatigues he incurred to maintain his political views in a great measure brought on the fatal illness which carried him off at Nice, the 23d of February, 1882."

The beginning of the controversy with Forbes is recorded in the following letter from Agassiz, addressed to Mr. Murray's son in London, who was the editor of the "Annals and Magazine of Natural History."

NEUCHÂTEL, 13 février, 1842.

D'après votre lettre je présume que c'est Forbes qui vous a offert un article sur les glaciers; si c'est lui ce serait une raison de plus pour moi de vous prier d'attendre mes notes, car je viens de recevoir une notice de lui insérée dans le Journal de Jameson(*Edinburgh New Philosophical Journal*) qui me paraît la plus complète indiscrétion dont on puisse se rendre coupable envers un ami. Mr. Forbes à mon invitation est venu passer trois semaines dans la cabane que j'avais fait établir sur le glacier de l'Aar, je lui ai fait voir tout ce que la glacier offre d'intéressant, toutes les recherches ont été suivies sous ses yeux. Dès le premier jour je lui ai même annoncé que l'un des points que je me proposais spécialement d'étudier était la structure intime du glacier et particulièrement les apparences rubannées du glacier que j'avais à peine mentionnées dans mon ouvrage ("Études sur les Glaciers"), p. 121, en en décrivant l'aspect extérieur, parce que depuis 1838 où je les avais pour la première fois remarquées sur la mer de glace de Chamounix je n'étais point encore parvenu à en suivre tous les détails, ce phénomène n'étant pas toujours également distinct. Comme cette année il a été facile de l'observer, nous en avons fait une étude détaillée et dès les 3 octobre, 1841, j'en donnai la description à M. de Humboldt, alors à Paris, qui en fit part à l'Académie (des Sciences de l'Institut de France), et voilà qu'en décembre (*Edinburgh New Philosophical Journal*, January, 1842), Mr. Forbes en fait part à l'Académie d'Edinburgh en s'en appropriant la découverte et en poussant l'impudence jusqu'à dire qu'il fut surpris en visitant le glacier de l'Aar de voir en me parlant de ce phénomène que je ne le connaissais pas. Veuillez à ce sujet comparer la page 121 de mon livre. Il faut absolument que je fasse connaître ces faits pour ne pas paraître plus tard plagiaire dans mes propres observations, et je vous prie de communiquer le contenu de cette lettre à tous ceux de mes amis que vous connoissez et que cela peut intéresser. Ceci est une raison de plus pour activer la redaction de mes observations sur les glaciers, et je compte que votre sentiment de justice vous engagera à les attendre plutôt que d'accepter quelque fausse monnaie. Je vous prie cependant de ne pas faire imprimer ceci parce que je compte faire moi-même la leçon à Mr. Forbes.

Je serais moins surpris de ce que vient de faire Forbes, si lorsque nous étions ensemble et que je le priais de contribuer à faire connaître avec moi les glaciers en prêtant à cette question l'appui du nom d'un physicien justement estimé dans le monde savant, il s'y était continuellement refusé, en répétant qu'il n'avait aucune opinion sur ce sujet, qu'il avait voulu seulement apprendre à les connaître en venant les étudier avec moi et qu'il se garderait bien de rien publier sur une matière dans laquelle il lui restait plus que des doutes J'étais bien loin de présumer que sous cette réserve se cachait l'intention de s'approprier les observations les plus précieuses de cette campagne.

Mr. Forbes a soin de dire que c'est dans ma société et celle de Mr. Heath de Cambridge qu'il a séjourné trois semaines sur le glacier. Pour être vrai il aurait dû dire que j'avais fait établir là haut un appareil de forage desservi par cinq ou six hommes y compris un maître foreur, que je m'étais fait accompagner d'un peintre qui a dessiné pour moi tous les accidents du glacier dont quelques uns ont été copiés par Mr. Forbes; que deux de mes amis, M. le Docteur Vogt m'aidait dans les recherches microscopiques et M. Desor dans celles concernant la géologie; c'était tout un établissement que Mr. Forbes s'appropie gratuitement par un prétentieux *our*. Qu'en serait-il si de leur côté les autres savants MM. Studer, Escher, Martins, etc., qui sont venus nous visiter et passer quelques jours avec nous, en faisaient autant?

Agassiz took the side of his secretary, and published, on the 21st of April, 1842, a pamphlet of ten octavo pages, without title, but which may be called, "A reply to Mr. James D. Forbes on the laminated structure of glaciers." The paper, although "privately distributed," was circulated largely among Agassiz's friends to the number of five hundred copies. It began with a reprint of a letter to M. Desor, under the date of 11th March, 1842, published by Forbes, with the remark: "The following letter from Professor Forbes to M. Desor of Neuchâtel was written in answer to one from the latter to the former, in which Professor Forbes is charged with having, in a paper on the structure of the ice of glaciers . . . assumed as his own a discovery previously well known to M. Agassiz and his friends. It appears that this injurious assertion has been pretty extensively circulated through private channels, and, in consequence, Professor Forbes has been advised by his friends to make his denial equally known." The assertion that Desor's written letter "has been pretty extensively circulated through private channels" involved a gratuitous supposition without foundation; the letter was not printed, and neither Agassiz nor Desor had even kept a copy of it. Forbes's printed letter, on the contrary, was largely circulated, although no copy was addressed to Agassiz; and Agassiz was obliged to make use of the "Private copy for Mr. Guyot" in order to have it reprinted in his pamphlet. If the "confidential adviser" of Agassiz, as Forbes calls Desor, erred in writing to Forbes in a rather surly tone, Forbes's letter is much more objectionable. In it he calls to his help Studer, who was always only too ready to join in a crusade against intruders on his geological preserve of the Bernese Oberland. Forbes seems rather anxious not to appear to have studied "in the school of Agassiz"; but to show that the fact of the structure of the ice described in his notice was unknown to Agassiz, de Charpentier, and other writers. His remarks about "studying in a school" are childish in the extreme, and his knowledge of other works on the structure of the ice was certainly limited; for Hugi, Rendu, and before them David Brewster of Edinburgh, had observed the veined structure.

Agassiz's answer, dated the 29th of March, 1842, gives the whole story of the relations between him and Forbes. There is no doubt that Agassiz and every one who met Forbes under the auspices of Agassiz, both at the "Hôtel des Neuchâtelois" and at Neuchâtel, did everything possible to help Forbes, and were extremely kind and courteous to him; while, on his part, Forbes was austere to an extent seldom seen, even among Englishmen. The impression he made when in Switzerland was decidedly unfavourable, except in the case of a single person, Professor Bernard Studer, to whom he afterward dedicated his book, "Travels through the Alps of Savoy," etc., 1843. It was wrong on his part to accept the hospitality of Agassiz, and then to act as if he had met him in a hotel. He was constantly on his guard not to show any mark of assent, or to say anything which might be useful for future observations. His great reserve puzzled everybody; and when he left, there was a general feeling of relief. Throughout his stay the relations were never cordial. He personified the celebrated type of Englishmen so well described and caricatured by Töpffer, in his "Nouvelles Genevoises," "Les deux-Scheidegg": "Je défendé vos de paaler à moa, quand je disé rien à vos." A true no! no![1] tall, thin, dry, haughty, and extremely egotistical.

Agassiz put forward the doubtful claim of Arnold Guyot to priority in the discovery of ribboned structure, noted by Hugi as far back as 1830. It would have been better had no attention been paid to Forbes's paper, which was written in bad taste and against all the rules of courtesy between savants.

The only person who obtained any benefit from this uncalled-for dispute was Desor, whose name, until then entirely unknown in England and on the continent, except in Switzerland, became conspicuous as the "confidential adviser" of Agassiz.

All friendly relations between Agassiz and Forbes ended with the following letter addressed by Agassiz to Forbes, then in a hotel at Neuchâtel:—

LE 12 JUIN, 1842.

Monsieur,—Je viens de recevoir la brochure que vous m'avez fait l'honneur de m'adresser et pour laquelle je vous prie d'agréer mes remerciements. Je regrette que vous n'ayez pas encore reçu le récit de notre course à la Jungfrau que M. Desor vous a adressé il y a plusieurs mois, si j'en avais encore un exemplaire à ma disposition je vous l'adresserais, afin que vous puissiez en prendre connoissance.

[1] Rudolph Töpffer in his novel "Le Col d'Anterne" givès, as a type of a well-bred Englishman, a tourist set in front of the Mont-Blanc, who disclaimed to answer any of the numerous and polite questions asked by Töpffer, except by the two words, No! no! and Ui! ui!

HOTEL
des
Neuchatelois
1840
E. DESOR
STUDER
ESCHER
C. Vogt
C. Nicolet
F. Pourtalès
H. Weber
Dollfus
F. Keller
A. FAVRE
J. E. Pictet

N'ayant recu aucune réponse aux deux dernières lettres que j'ai eu l'honneur de vous adresser et après la réponse que vous avez faite à ma précédente, en livrant au public des remarques qui n'allaient qu'à votre adresse, je ne conçois pas quelle espèce de relations personnelles vous pouvez rechercher avec moi. Celles que j'aurais pu désirer, vous les avez rendues impossibles; et je ne saurais accepter les froides civilités d'une personne en qui j'avais vu un ami. Cela ne m'empêchera pas de rendre pleine et entière justice à celles de vos publications qui tiennent de loin ou de près aux recherches scientifiques que je poursuis.

Agréez, Monsieur, etc.,

LOUIS AGASSIZ.

It would have been wiser on the part of Agassiz and more profitable if, after his ascent of the Jungfrau and his two "séjours" at the "Hôtel des Neuchâtelois" in 1840 and 1841, he had let the glacial question take care of itself. The impulse he had given was quite sufficient to assure his reputation as one of the first and most successful workers; and his place, after Venetz and de Charpentier, was recognized as undisputed by all those who had studied glaciers and the glacial age.

Frightened at the constant increase of expenses, his Swiss and German families made remonstrances, and were absolutely opposed to a new estáblishment at the glacier of the Aar, to replace the "Hôtel des Neuchâtelois," which had gone to pieces during the winter, according to a report just received from the Grimsel. Agassiz's best scientific friends, with Humboldt at their head, hinted that, after all, his works on fishes furnished his best claim to reputation and celebrity. In a previous letter dated Berlin, 17th of June, 1838, Humboldt, in a friendly way, had told him that he had never had a secretary, or even a copyist, preferring to do all his writing, and expressing his fear when he learned that he had nine assistants in his employ, adding humorously, "I am sure that there must be some gold somewhere in your polished rocks. I should like to know your secret how to work so profitably and so quickly all these mines." Humboldt repeated his friendly advice during the summer of 1842, saying plainly that he, the man of the equinoctial region, was frightened by the *Eiszeit* and the terrible ice cap. But all this was in vain. Agassiz had an answer for every objection; and all that even his alarmed mother could obtain was a promise that he would not make any more ascents of inaccessible peaks, and be lowered again into hell, — "descente aux enfers," as his descent into the glacier well was familiarly called.

Arrived at the Grimsel, the 7th of July, 1842, Agassiz, with his numerous assistants, at once began observations and excursions, first to the Siedelhorn, and after that to the glacier of the Rhone. The troglodytic habitation under the immense block, having become unsafe, it was replaced by a long tent, divided into three compartments, used as laboratory and dining-room, sleeping-room, and dormitory for the workmen. The form of the tent — twenty metres long, four metres broad, and five metres high — recalled a Noah's ark, and was therefore christened "the Ark,"[1] to distinguish it from the "Hôtel des Neuchâtelois," which was now used as a kitchen. The old cabin of the guides served as a stable

[1] The old name of "Hôtel des Neuchâtelois," however, continued in use; and the archaic word "Ark" was dropped before the summer was over.

for ten goats; and the establishment, as a whole, was a great improvement on the old one. Besides being built on the solid ground and not on the moving median moraine, it afforded a shelter, beneath which they could work whenever the rain obliged them to keep in doors. On the 10th of July it was ready for occupancy; and the same night Agassiz, Wild, Vogt, Nicolet, Desor, Burkhardt, and Girard slept under the canvas-covered cabin. A new member was added to the usual staff of Agassiz, — M. Wild of Zürich, who had been engaged by Agassiz as a topographical engineer, to survey and make a trigonometric plan of the Aar glacier. To be sure, the king of Prussia, on the recommendation of Humboldt, had granted nearly a thousand dollars for the continuance of Agassiz's glacial work; but this royal gift was soon expended, and before the campaign of 1842 was over, Agassiz was more deeply in debt than ever; for with him a gift, however large and important, was only an occasion to expend twice and three times more than he had received.

The stay at the glacier extended from the beginning of July until the middle of September, with numerous excursions, one as far as Altorf to attend the meeting of the Helvetic Society of Natural Sciences. Escher de la Linth and Ferdinand Keller (the same who twelve years after made the first discovery of the lacustrine habitation of prehistoric man) were among the guests who helped to make observations and experiments on the glacier. Numerous other guests came, but only as visitors and spectators. Investigations were made in regard to infiltration, lamellar structure, stratification of the glacier, the purity and composition of the ice, the "crevasses," the temperature, the motion of the glacier, the ablation, and the *névé*. Agassiz had resolved to embody in a large publication, in three parts, everything relating to the glacial system. The first part, the only one ever published, was entitled "Nouvelles études et expériences sur les glaciers actuels, leur structure, leur progression, et leur action physique sur le sol," and was accompanied by a beautiful folio atlas, containing three maps and nine plates (Paris, 1847). The second part was to be on the "Alpine erratics," by Guyot. It was never written, only a few general conclusions being published, without maps of any sort. It is to be regretted, for Guyot had prepared a map showing the distribution of the Alpine boulders, which had not been published. However, a great part of it — more than two-thirds at least — was anticipated by the issue, in 1845, at Winterthur, of an anonymous map of the old glaciers of the Swiss Alps, showing the extent of the ancient glaciers of the Arve, Rhone, Aar, Reuss, Linth, and Rhine, with their lateral and frontal moraines. This map is entitled "Verbreitungsweise der Alpen-fündlinge," and its author is the modest and very able geologist, Arnold Escher von der Linth. Very likely the publication of this map discouraged Guyot, who was always extremely slow and timid; and he resolved to publish neither the volume advertised nor his map. As to the third and final part, by Desor, treating of erratic phenomena outside of Switzerland, it remained in the stage of contemplation, and was never begun.

Agassiz and all his collaborators and friends certainly worked hard and with a determination to penetrate all the secrets of the glaciers, and some of their observations and experiments are excellent and valuable; but it is no injustice to any of them to say that they were not sufficiently equipped and prepared for the work they had rather rashly undertaken. Devotion to progress of science was not sufficient; something more was required. De Charpentier and Bishop Rendu had already said all that could be expected from men not trained as physicists. Agassiz added very little, if any, to their work. What was wanted was a great physicist, to solve the problem of glaciers. James D. Forbes proved unequal to the task; and it was reserved for John Tyndall, the great pupil and successor of Faraday, as the discoverer of "radiant heat," to explain fully the origin of glaciers, the pressure theory, regelation, crystallization and internal liquefaction, the veined structure; in fact, all the mechanism of glaciers. The principles set forth in Tyndall's "The Glaciers of the Alps" (London, 1860), come next to the great discoveries of Venetz and de Charpentier, and to Agassiz's Ice-age. The four complete the survey of the subject.

In November, 1842, Agassiz, losing patience with the constant attacks in German newspapers directed against him by his formerly intimate friend, Dr. Karl Schimper, published a pamphlet entitled "Erwiederung auf Dr. Karl Schimper's Angriffe," four-page quarto, for private circulation, though it was freely distributed, more especially in Germany and Switzerland. It would have been better Agassiz had ignored these attacks; but urged on, he says, by many friends, and I may add by the one called by Forbes "his confidential adviser," he wrote his "Reply to Dr. Karl Schimper's Attacks." In it, interesting details of their life as students, and of the sort of community existing at that time between Alexander Braun, the two Schimper brothers, and Agassiz are given, and the kindness and generosity of Agassiz to the two Schimpers are revealed; full justice is done to the brilliant intellect of Karl Schimper, and his share in the diagram entitled "Crust of the Earth as related to Zoölogy," constructed by him with the help of notes furnished by Agassiz, and afterward published (1848), is fully acknowledged. As to Agassiz's delay in returning specimens of fishes lent to him for his great work on fossil fishes, it was unavoidable, on account of the many specimens to be taken care of, and the delay in the publication. As soon as the work was finished, every specimen was carefully packed and returned in good condition.

Schimper's claim to a small collection of minerals offered to Agassiz at Carlsruhe, when Agassiz was on the point of beginning his lectures as professor at Neuchâtel, shows only too plainly how depressed and demoralized Schimper had become after the break in his relations with Agassiz in 1838.

The only fault, and it is a very trivial one, to be found with Agassiz, is that he did not refer to Schimper again in his volume "Études sur les Glaciers," in regard to the otherwise erroneous explanation of the diminution of the temperature of the globe with the disappearance of the animals, analogous to the phenomena accompanying the death of individuals, and then of its rising again, due to the arrival of a new creation of animals, developing heat as a consequence. In his volume Agassiz reproduced Schimper's small mathematical figure, and it would have been well to quote Schimper as his authority. Alexander Braun, when consulted, threw the blame on Agassiz, but refused to take part in the dispute. In a letter from Agassiz to Braun, published in Braun's Life, by his daughter, he says that if he did not quote and speak of Schimper in his "Études sur les Glaciers," it was in order to punish Schimper for his unjustifiable conduct towards him; a very lame excuse, for scientific ideas and discoveries are sacred property, which cannot be cancelled under any circumstances. If Agassiz had repeated the sentence in his Neuchâtel Address of 1837, "l'explication de tous ces phénomènes (glaciaires) est le résultat de la combinaison de mes idées et de celles de M. Schimper," everything would have been satisfactory.

It is strange that Agassiz did not abandon the theories advanced in his "Discours de Neuchâtel," after its delivery; for they met with not the smallest acquiescence or encouragement, either from those who heard the address or from those who read it afterward. De Charpentier was against it, and Sedgwick, the celebrated geologist of Cambridge (England), expressed in happy terms the impression made on him by the reading of the "Études sur les Glaciers," when he said: "I have read his Ice-book. It is excellent, but in the last chapter he loses his balance, and runs away with the bit in his mouth."[1]

[1] "Life and Letters of Sedgwick," Vol. II., p. 18, Cambridge, 1890.

The immense "nappe" of ice covering the earth, its breaking by the upheaval of the Alps, etc., seem the theoretical views of a dreamer, and are entirely at variance with the excellent and remarkable observations on the power of glaciers to carry boulders, and their great extension during the Quaternary epoch. But it was a special characteristic of Agassiz's mind, which was intensified by the teaching of his great master Cuvier, seldom to acknowledge an error, but on the contrary to try by all means to maintain his position. He repeatedly made mistakes in dealing with other savants, and also in the too hasty generalizations which he sometimes put forward in natural history. I do not hesitate to attribute these weak points in Agassiz's character to the influence of the author of the "Anatomie Comparée," an influence which, if profitable on many accounts, was sometimes much to be regretted. At all events, Cuvier's influence was profound, and among many things that Agassiz learned in his laboratory, was one of his most pronounced faults, the authoritative and tyrannical attitude of the master, unable to accept a contradiction, or to abandon an idea, when once promulgated and in print.

The polemic with Karl Schimper was unfortunate, because Schimper was no longer responsible. Like all persons suffering from mental disorder, he thought he had discovered all that he had heard in regard to the glaciers and the glacial question during his long visits at Bex and at Neuchâtel, and he treated very slightingly Venetz, Charpentier, and Agassiz; he attributed to himself the lion's share, when he was only a poetical echo, and a rather fantastic one at that, of what he had heard during his stay in Switzerland. His half-scientific, half-humorous poem "Die Eiszeit," printed at Neuchâtel, for friends, the 15th of February, 1837, the birthday of *Galileo Galilei*, whose name Schimper had assumed when a student, shows the state of mind into which he had already sunk; that of an obscured spirit. Schimper, after brilliant "débuts" in science, produced nothing but two small volumes of indifferent poetry, entitled "Gedichte" (Erlangen, 1840, and Mannheim, 1847); he published nothing on the morphology of plants, although he is justly regarded as one of its discoverers. Charged by Prince Maximilian of Bavaria, in 1842, to make a survey of the Bavarian Alps and the Palatinate, he made no report, and finally was confined in an asylum, at Schwetzingen, where he died the 21st of December, 1867.

There is no doubt that Schimper was a well gifted man. Without publishing a word, he left, as a botanist, a reputation of a high order, and he influenced both Alexander Braun and Agassiz to a great extent, possessing more imagination and original ideas than either of them. "Il n'a manqué à Schimper que d'être sobre," one of those who knew him best once said to me.

Among the visitors attracted by curiosity to the "Hôtel des Neuchâtelois," during the summer of 1842, was a great manufacturer of Mulhausen, M. Daniel Dollfus-Ausset. Such an enthusiast of high regions, of glaciers, and of the glacial question, has rarely existed. He was so fascinated by all that he saw on the glacier of the Aar, that from that day he became not only an adept, but one of the most generous patrons of the work in progress on the Oberland glaciers. As a first step, he begged Agassiz to accept for him, and as many of his assistants as he wished, an invitation to pass the week between Christmas and New Year with him at the Hôtel des Trois-Rois at Bâle, as a relaxation from their hard work, and to celebrate his enrolment among the glacialists, and the inhabitants of the "Hôtel des Neuchâtelois." Agassiz, in company with Desor and Vogt, left Neuchâtel the 24th of December, 1842, and arrived at Bâle in time to celebrate "Le réveillon," or Christmas Eve. "Papa Dollfûs," as he was always called afterward, received them most cordially, and for a whole week, with the exception of daily morning work at the Museum of Natural History, under the direction of the learned and very sociable Peter Merian, they were treated as princes of the sciences.

And thus Agassiz and his assistants ended the year 1842 at Bâle, in the enjoyment of a royal hospitality.

CHAPTER IX.

1843–1844.

"RECHERCHES SUR LES POISSONS FOSSILES," 1833–1843 — REVIEW OF IT BY JULES PICTET DE LA RIVE — DR. A. GÜNTHER'S OPINION — AGASSIZ'S ERRORS WITH THE EOCENE FOSSIL FISHES OF GLARIS (SWITZERLAND) — THE PART TAKEN BY COLLABORATORS IN THE "POISSONS FOSSILES" — ANOTHER VISIT TO THE GLACIER OF THE AAR — THE MEETING OF THE HELVETIC SOCIETY AT LAUSANNE, JULY, 1843 — AGASSIZ'S HOSPITALITY AT NEUCHÂTEL — FALSE POSITION OF HIS SECRETARY, DESOR, AND HIS ASSISTANT, VOGT — SCIENTIFIC LIFE IN NEUCHÂTEL — "MONOGRAPHIES DES POISSONS FOSSILES DU VIEUX GRÈS ROUGE," 1844 — THE GEOLOGIST AND STONECUTTER, HUGH MILLER — "HISTOIRE NATURELLE DES POISSONS D'EAU DOUCE" — KARL VOGT LEAVES AGASSIZ — EXTRAORDINARY SESSION OF THE GEOLOGICAL SOCIETY OF FRANCE AT CHAMBÉRY (SAVOY) — FAILURE OF NICOLET'S LITHOGRAPHIC ESTABLISHMENT — DINKEL LEAVES NEUCHÂTEL — ILLNESS OF GRESSLY.

THE publication of the "Recherches sur les Poissons fossiles" continued through the ten years from 1833 to 1843, when the eighteenth and last part or "livraison" was issued, with "Additions à la préface," dated Neuchâtel, May, 1843, which may be considered as the last pages of that great work. It is a true monument to the science of palæontology, and to speak of it with authority requires such special study of ichthyology, that the only way to give an idea of its value is to quote one of the very few men able to speak of it with "connoissance de cause." For this reason, I have chosen to quote what Jules Pictet de la Rive says of it in his article, "Agassiz" ("Album de la Suisse Romane," 5ième vol., Geneva, 1847) Pictet had made a special study of fossil and living fishes, and his intimacy with and admiration for Agassiz never relaxed during his whole life. Independent by character and possessing a large fortune, Pictet's opinions are properly considered just and unbiassed.

"The 'Recherches sur les Poissons fossiles,'" says Pictet, "was one of the first conceptions of Agassiz, and form to-day his most substantial title to renown. It is in this beautiful work that the immanent qualities of our learned palæontologist shine more specially and that his rich imagination has full play, although always guided by a sagacious and well-balanced judgment based on conscientious researches and on a minute analysis of even the smallest parts of the organism.

"The limits of this article do not allow us to give a complete idea of the work, which is composed of five quarto volumes and a folio atlas of almost four hundred plates. We shall only try to set forth the aim, the plan, and the most important results.

"We know that when Cuvier published his first works on the fossils his principal aim was to demonstrate that the species destroyed by the revolutions of the globe and preserved as fossils are different from those living now on our continents and in our seas. That truth has to-day become unquestionable, and new discoveries have shown by the most undeniable evidence, that there have been in the history of the earth a series of epochs during which the forms of the oceans and the continents have been successively modified, and each one of them has been characterized by a special flora and fauna; that is to say, by an ensemble of vegetables and animals specifically different from those coming before or after. The fishes have existed since the oldest ages of the globe, and their remains are found in all the successive periods. Their palæontological history, consequently, is most important, and furnishes precious data concerning this succession of faunas.

"When Agassiz began his researches and foresaw the importance of the result that he might draw from them, the classification of fishes was not advanced enough to allow sufficient comparisons. Some dissimilar forms were associated together, while other very similar ones were separated by large intervals. Before everything else, it was necessary to establish an exact classification. Agassiz found in the scales the necessary elements to solve the problem, and he recognized that these teguments of the body correspond well with the interior characters, and that their variations are, in general, associated with and due to, organic differences. He accordingly divided fishes into the following four orders: (1) The Cycloids, with scales, rounded, smooth, and simple at the margin, composed of laminæ of horn or bone, but without enamel,—endo-skeleton ossified; (2) the Ctenoids, with scales jagged or pectinated (like the teeth of a comb) on the posterior margin, formed by laminæ of horn, but without enamel,—endo-skeleton ossified; (3) the Ganoids, with angular scales regularly arranged like paving-stones, and composed of horny plates covered with a strong, shining enamel,— endo-skeleton cartilaginous, in some partly osseous and partly cartilaginous; (4) the Placoids, with cartilaginous skeleton and skin covered irregularly with enamel plates, sometimes of considerable dimensions, at other times reduced to small points like the prickly, tooth-like tubercles on the skin of rays.

"This classification allowed easy comparisons and generalizations, and the palæontologic history of the fishes offered results not at all expected and most important. These animals have been completely renewed by successive creations, and whole populations of them have been destroyed to make room for others which were very different. Of the four orders indicated above, the Placoids, or cartilaginous fishes, have existed during all the geologic periods, though they have undergone various modifications, most remarkable especially in the teeth. But the other three orders — that is, the osseous fishes — have somehow replaced one another. Our present seas contain almost altogether Ctenoids and Cycloids, and, except two genera of Ganoids living in rivers of warm countries, these two orders compose all the present fauna of osseous fishes, while, on the contrary, none existed before the deposit of the chalk, and it would be vain to look in all the preceding epochs for one Ctenoid or one Cycloid; that is to say, the old seas did not contain a single fish with thin horny scales like our perches or our trout, while in the present fresh waters and seas we find such fishes almost altogether.

"On the other hand, the Ganoids were most common previous to the Cretaceous epoch, and that order, now reduced, as I have said above, to only two genera, then formed the majority of the population of the seas. These same fishes present in their history a very remarkable fact. Until the Lias epoch, all the Ganoids possessed on the superior part of the tail a lobe formed by a prolongation of the vertebrate column. But from the Lias, on the contrary, all had a tail formed as that of the osseous fishes of the present time; that is to say, the vertebrate column stops at the base of the tail.

"It is, therefore, possible to divide the palæontological history of the fishes into three periods or epochs. During the first, extending from the Silurian to the Trias, the faunas are composed of Placoids and Ganoids with the vertebrate column prolonged to the upper lobe of the tail. In the second, which corresponds to the Jurassic epoch, we find the Placoids and Ganoids, with the ordinary tail. In the third, which began during the Cretaceous epoch and continued in our modern period, the Placoids, the Ctenoids, and the Cycloids form almost entirely the ichthyological population of the world. Hence, if a geologist found a Ganoid with a prolonged tail, he could conclude that the strata in which he found it belonged to the first period; a Ganoid with an ordinary tail would indicate, with sufficient certainty, that the group of strata belonged to the second epoch; and so on.

"It is easy to understand the interest created by a work, the aim and result of which are to demonstrate such remarkable laws, more especially when the proofs are based on an incredible number of facts and observations. The work of Agassiz mentions more than one thousand fossil fishes, with descriptions and beautiful plates, which make them known almost as well as if we were able to observe them alive.

"This work brought its author complimentary distinctions from several academies and learned societies. Particularly during a journey in England and Scotland all the collections were open to him, assistance in various ways was offered, and he had the great satisfaction of seeing with what astonishing precision the numerous new facts which he daily observed confirmed all his previous conclusions. The English and Scotch geologists for many years kept the remembrance of some keen anecdotes on the subject.

"Agassiz's researches opened a new path, through which he continued to advance, publishing in the meantime supplements to his main work; among them, a 'Monographie des Poissons du vieux grès rouge,' which was soon followed by one on the 'Poissons de l'Argile de Londres.' The first of these monographs furnished some interesting results, both geological and zoölogical; in particular, it demonstrated two most important laws: 1st, the analogy existing between the first condition of the embryos of fishes and the organization of fossil fishes of the oldest epochs; 2d, the parallelism existing between the embryologic development of the fishes and the succession of the different types of these animals in the series of formations."

There is nothing to add regarding the great value of this "vaste publication," as it is called by Pictet; but a few words are necessary to indicate some of the criticism which it called forth, and to meet claims which have now and then been put forward.

Agassiz knew perfectly well that his classification was artificial, and not based on all the natural principles, as it should have been, and as Cuvier's was before him; but he wanted to make use of a great quantity of fragmentary specimens, and even mere scales of fishes, which were found in abundance, and which otherwise would have been useless, and would have left a great gap in his series of forms. He worked as much to prove the succession of fishes in the different systems of strata, as to obtain a knowledge of them zoölogically, trying to find laws which might be used in palæontology to classify groups of strata by their fossil fishes. And he succeeded admirably, notwithstanding the defect of his empiric classification.

As Dr. A. Günther says: "We have no hesitation in affirming that if Agassiz had had an opportunity of acquiring a more extensive and intimate knowledge of existing fishes before his energies were absorbed in the study of fossil remains, he would himself have recognized the artificial character of his classification. The distinctions between Cycloid and Ctenoid scales, between Placoid and Ganoid fishes, are vague, and can hardly be maintained. So far as the living and post-Cretacean forms are concerned, he abandoned the vantage-ground gained by Cuvier; and therefore his system could never supersede that of his predecessor, and finally shared the fate of every classification based on the modifications of one organ only. But Agassiz has the merit of having opened an immense new field of researches by his study of the infinite variety of fossil forms. In his principal work, 'Recherches sur les Poissons fossiles,' 1833–1843 (4to, atlas folio), he placed them before the world, arranged in a methodical manner, with excellent descriptions and illustrations. His power of discernment and penetration in determining even the most fragmentary remains is truly astonishing; and if his order of Ganoids is an assemblage of forms very different from what is now understood by that term, he was, at any rate, the first who recognized that such an order of fishes existed."[1]

Agassiz was one of those naturalists who find it easier to discover differences than to bring together specimens of fossils. He possessed a rare power of discerning the smallest differences between allied forms of animals; but sometimes he went too far, as in the case of the Eocene fossil fishes in the flysch of Glaris (Switzerland), where the cleavage resulting from the breaking and compressing of the strata, during the dislocation of the Alps, deformed some specimens to such an extent that Agassiz was led to establish six species of *Anenchelum*, all of which really belong to a single species, *Lepidopus glaronensis*. The same mistake has been noted by Dr. A. Wettstein and A. Heim for species of the genus *Palæorhynchum*, *Acanus*, etc. ("Actes de la Société Helvétique des Sciences naturelles," Geneva, August, 1886, pp. 46, 47). India-rubber models of some of these fossil fishes, when pulled in certain directions, give as many species as Agassiz founded; and it is evident that Agassiz, in some cases, too easily multiplied

[1] "Ichthyology," by A. Günther, in "Encyclopedia Britannica," ninth edition, Vol. XII., p. 634, London, 1881.

the number of species without proper restriction. But this is only a detail, which does not affect the final result and conclusions, nor the prodigious capacity of his memory, in which lay the true secret of his classification of fossil fishes.

In regard to the help that Agassiz received in his "Poissons fossiles": in the first place, the excellent drawings were made by Dinkel and Mrs. Agassiz, those of the latter being fully as good as and rivaling in execution the best of the artist Dinkel. Secondly, after the issue of the first twelve parts or "livraisons," Agassiz made a great deal of use of his assistant Karl Vogt and his secretary Desor, in preparing the bones and the scales, and in writing the descriptions of species and even of genera. But as Vogt wrote me: "Agassiz avait parfaitement le droit de s'attribuer ces travaux, car il me payait pour cela, j'étais son préparateur à gages sous ce rapport." Only one-third of the work was thus prepared with his two collaborators, under Agassiz's direction; but this may be said, that it would have been much better if he himself had finished what he had so well begun and continued until 1838.

At the end of July, 1843, Agassiz returned to his work on the glacier of the Aar. A new cabin had been erected, which was called the "Pavillon"; and Daniel Dollfus-Ausset, with his son, established himself close by, in another cabin. The time was passed in measuring the motion of the glacier, its temperature, etc., and in Alpine climbing. On the whole, it was a rather expensive campaign, and the results were inadequate compared with the money expended.

During the meeting of the Helvetic Society of Natural Sciences at Lausanne, the 25th and 26th of July, Agassiz made a verbal communication in regard to his researches on the glaciers, speaking of the new and more practical direction given to his studies, and insisting on the stratification of the glaciers and the blue bands of ice, and on the formation of crevasses. At the same meeting, he spoke of the great value of fossil fishes in determining the ages of the "terrains," and more particularly of the squaloid teeth, like those of the true sharks, or squalodonts, the *Ptycholepis* of the Chalk, the *Strophodus* of the Jura, the *Acrodus* of the Lias, and the *Psamnodus* of the Coal measures.

His great sociability, which attracted so many people to the "Hôtel des Neuchâtelois," was exercised on rather a large scale in Neuchâtel, if we may judge by the following letter: —

NEUCHÂTEL, 23 décembre, 1843.

PROFESSEUR JULES PICTET DE LA RIVE,
Genève.

Mon cher ami, — Favre vous aura fait part du désir que j'ai de réunir ici quelques amis Jeudi prochain. Je viens insister auprès de vous pour que vous soyez de la partie.

En arrivant Mercredi soir et en descendant "Aux Alpes," vous trouverez Mérian, Escher, Studer et Valentin.

L'un de mes guides m'a procuré un jeune chamois dont nous dépécerons les os Jeudi chez moi. Faites-moi le plaisir de venir; si vous pouvez, amenez les Plantamour.

Votre tout dévoué,

LS. AGASSIZ.

His prodigious power of attraction is shown in his ability to bring together, for a single dinner party, at

Christmas time, during the snowy season, savants from every part of Switzerland, from Bâle, Zürich, Berne, Geneva, etc., and at a time when travelling was not easy, as it is now, with railroads in every direction. Nothing shows better that Agassiz was an accepted leader among the scientific men of Switzerland.

The year 1844 was a sad year with Agassiz. We must turn back a few years in order to understand the state of affairs, and how, little by little, he jeoparded his position by a complete incapacity to manage his assistants, his many employees, and his too numerous undertakings. Too great familiarity with his assistants, and inability to keep them at respectful distance, resulted in his having no authority over them. If Agassiz was a genius in natural history, in private life he was entirely unable to manage his immediate surroundings. Speaking of Agassiz's establishment at Neuchâtel, Karl Vogt says: "It was a scientific factory with a community of property; only, unhappily, neither the number of workmen nor the capital engaged was sufficient and in proportion to the production." It was also an overworked establishment. Agassiz, as its director, had to provide everything; first the money, for all were penniless; and the life they led, though without luxury, was, after all, rather expensive; for to travel all over Switzerland, to stay at the "Hôtel des Neuchâtelois," to keep open house at Neuchâtel not only for his assistants, but also for all the naturalists who were continually coming from every part of Europe, required a constant expenditure of no small amount of money. Besides the work of providing the money, Agassiz had an oversight of every work going on; he had to dictate letters, to insert sentences in the descriptions of his assistants in order to connect them and give them unity, to read and correct at least one of the proofs; even to direct the draughtsmen, and to select the drawings to be used, in regard to the artistic merits of which he was very critical and a capital judge, seeing faults where others were glad to admire the fine execution. Agassiz was well seconded by the artists in his service; but scientifically the assistance he received was rather deficient. Karl Vogt had been educated as a naturalist, and soon became most efficient in regard to the anatomy and the embryology of fishes; he also worked out the osteology and neurology, prepared the specimens, made the drawings, and wrote the descriptions. He was a first-rate assistant, knowing well his duties; and during the five years of his connection with Agassiz he did a great amount of good work. Although he always insisted that he was not a pupil of Agassiz, having learned zoölogy in Germany, there is no doubt of the great influence exerted by Agassiz on his work during the first ten years of his life as a naturalist.

In October, 1837, as we have seen, Agassiz engaged Desor as his private secretary, who, until then had done nothing in natural history, with which he was not even acquainted, beyond the general knowledge possessed by any student of a university. Employed first as a translator and a writer of dictated letters, he soon acquired sufficient knowledge of fossil fishes and fossil echinoderms, to help in describing species. Under Agassiz's teaching he made such rapid progress that, in three years, he became a useful assistant, not only in palæontological works, but also in the work on the glaciers and the glacial question. Vogt says of him, that in 1840 Desor was the "cheville ouvrière" (keystone) of the whole Agassiz establishment; and Agassiz, on the 11th of June, 1840, writes:—

Dans la rédaction de cette seconde partie (Cidarides) de mon mémoire ("Description des Echinodermes fossiles de la Suisse") j'ai été continuellement assisté par M. Desor, qui a continué à me prêter l'appui de sa plume facile, comme il l'avait déjà fait pour la première partie. Mais cette fois son travail ne s'est pas borné à une simple rédaction; l'examen comparatif des nombreuses espèces des genres Diadème et Cidaris, dont les caractères sont si difficiles à apprécier, est même entièrement de son fait. Cependant j'en ai revu la description, afin d'en partager avec lui la responsabilité scientifique. Il m'est précieux d'avoir trouvé dans un ami un collaborateur aussi distingué.

Desor had no initiative faculty, and was totally devoid of original ideas. He never rose above a third-rate naturalist, retaining all his life the spirit of a lawyer, with a special tendency to politics and a politician's methods. Charles Girard was in too modest a position to be helpful scientifically, except in the work of compilation, which he always performed very industriously. As regards Gressly, the help he gave Agassiz was invaluable; the exact geological position of two-thirds of the fossils described in the different palæontological works of Agassiz was learned from him; and he furnished more than half of the best specimens of the echinoderms, the Myas and the Trigonias. In the scientific association directed by Agassiz, Gressly acted as the St. Bernard dog, faithful, true, living, no one knew exactly how, on the crumbs from the table always spread in Agassiz's home; always satisfied, always respectful, and never so happy as when Agassiz expressed his admiration of the beautiful and rare fossils he drew from his numerous and large pockets on his return from his never-ending explorations in the Jura. However, an assistant as modest and inexpensive as Gressly is a rare exception, and Agassiz never again found one like him.

It was evident that something was wrong in the whole establishment, and that running on such a basis it would not last long. In fact, 1844 was its last year, as we shall see further on. But before relating the numerous incidents which one by one occurred and finally destroyed, at least partially, that extraordinary and brilliant scientific centre, due entirely to the genius of Agassiz, it is pleasing to call attention to two of the best works done at Neuchâtel under the impulse of this remarkable man.

One of his most important works, and certainly his most original, is the "Monographie des Poissons fossiles du Vieux Grès rouge ou Système Dévonien [Old Red Sandstone] des îles Britanniques et de Russie" (4to, with a folio atlas of forty-one plates), which was issued by "livraisons," or parts, the last three being distributed in August, 1844. The material used was mainly the specimens collected at Cromarty, in the North of Scotland, by the celebrated geologist and stonecutter Hugh Miller.

During the ten years previous to Agassiz's visit at Cromarty, in September, 1840, Miller, with great patience and skill, had unearthed from the old red sandstone the most wonderful forms of animals yet found. Agassiz says of some of them: "It is impossible to see aught more bizarre in all creation than the *Pterichthyan* genus: the same astonishment that Cuvier felt in examining the *Plesiosaurus*, I myself experienced, when Mr. H. Miller, the first discoverer of these fossils, showed me the specimens which he had detected in the Old Red Sandstone of Cromarty." As early as 1831, Miller found the *Pterichthys*, or winged fish; but Agassiz did not hear of it until 1838, when a description and drawing was shown him in Paris by an English naturalist: he was greatly interested in this new form of life, and very anxious to see more of it. The following extract from Hugh Miller's principal and most popular work, "The Old Red Sandstone," explains how Agassiz was first made acquainted with Miller's wonderful discoveries:—

A letter which I wrote early in 1838 to Dr. Malcolmson, then at Paris, and which contained a rude drawing of the *Pterichthys*, was submitted to Agassiz, and the curiosity of the naturalist was excited. He examined the figure, rather, however, with interest than surprise, and read the accompanying description, not in the least inclined to scepticism by the singularity of its details. He had looked on too many wonders of a similar cast to believe that he had exhausted them, or to evince any astonishment that geology should be found to contain one wonder more ("The Old Red Sandstone" by Hugh Miller, p. 119, Boston, 1854).

Although Agassiz had great sympathy and very cordial relations with Hugh Miller, their correspondence was extremely limited. Mrs. Agassiz says that with a single exception no letters have been found from him among Agassiz's papers; and she gives that unique letter in Vol. II., pp. 470–477 of her work ("Louis Agassiz, His Life and Correspondence"). Lately another letter has been found in Switzerland by M. Auguste Mayor, and I here give an extract from it. The principal part is descriptive of specimens of fossil fishes sent to Agassiz, which would be unintelligible without good figures, and is consequently omitted; but the parts given are interesting on account of the great originality and keenness of the writer.

CROMARTY, 30th May, 1838.

PROFESSOR AGASSIZ,
Neuchâtel.

Honored sir, — I have just learned from my friend Dr. Malcolmson that you have expressed a wish to see one of the fossils of my little collection. I herewith send it you and a few others which you may perhaps take some interest in examining.

I fain wish I could describe well enough to give you correct ideas of the locality in which they occur. Imagine a lofty promontory somewhat resembling a huge spear thrust horizontally into the sea, — an immense mass of granitic gneiss, forming the head and a long rectilinear line of Old Red Sandstone the shaft. On the south side are the waters of the Moray Firth, on the north those of the Firth of Cromarty. The claystone beds which contain the fossils occupy an upper place on the sandstone shaft, covering it saddlewise from firth to firth. A bed of yellowish stone about sixty feet in thickness lies over them, except where they are laid bare by the sea, or cut into by two deep ravines — a bed of redder stone of unascertainable depth (though it may be measured downwards for considerably more than one hundred yards) lies beneath. The beds themselves average from ten to thirty feet in thickness. They abound everywhere in obscure vegetable impressions and fossil fishes, but in some little spots these last are much better preserved than in the general mass. All my more delicately marked fossils have been furnished by one little piece of beach hardly more than forty square yards in extent.

Of all the fossils of these beds, the one with the tuberculated covering seems least akin to anything that exists at present. I have split up many hundred nodules containing remains of this animal, for in the time of the Old Red Sandstone it must have existed by myriads in this part of Scotland. The larger ones I have invariably found broken and imperfect. The nodules in which they occur are in general too small to contain more than detached parts of them when large; and besides, the coat of the creature, consisting of hard plates separated apparently by sutures, must have offered a very unequal degree of resistance to the superincumbent weight. And, however, though the plates themselves are often as well defined and entire as the bits of a dissected map, they are almost always found displaced and lying apart. It is only the smaller fossils that I find perfect enough to furnish me with anything like adequate ideas of the original shape of the animal; but in these, though the general outline be better preserved, the plates are comparatively obscure. Thus the bits of the dissected map still want a key, and I have not yet become skilful enough to place them together without one.

The form of the body of the creature seems to have somewhat resembled that of a tortoise. . . . Pardon me, honored Sir, that I use this minute in describing these differences to you who observe better than any one else and can make a better use of what you observe. I have not succeeded in convincing some of our northern geologists that we have two varieties of small scaled fish in our beds, and I am now appealing to you as our common judge, and thus showing the ground of my appeal. Besides, as I cannot send you my specimens by hundreds, I deem it best (though it may seem presumptuous in one so unskilled) to communicate in this way the result of my examinations of the whole. One single specimen sometimes furnishes a characteristic tract regarding which perhaps fifty illustrations of the same fossil may be silent. Among all my specimens of the fish with the spines, only one shows me that the animal was marked by a lateral line. . . . I am afraid, however, that when thus communicating the results of some of my petty observations, I am but gaining for myself the reputation of being a tedious fellow.

I need not say how heartily welcome you are to the specimens I send you, should you have any wish to retain them. . . . Do I ask too much, honored Sir, when I request a very few lines from you to say whether the formation in which these fossils occur be a freshwater one, or otherwise, and whether the small scaled fish with the teeth be of a kind already known to geologists or a new one? I am much alone in this remote corner — a kind of Robinson Crusoe in geology — and somewhat in danger of the savages who cannot be made to understand why, according to Job, a man should be "making leagues with the stones of the field." But I am sanguine enough to hope that the good nature, of which my friend Dr. Malcolmson speaks so warmly, may lead its owner to devote a few spare minutes to render these leagues useful to me.

I am, I trust, sufficiently acquainted with geology, rightly to value the decisions of its highest authority.

I am, honored Sir, with sincere respect,

Your most obedient Servant,

HUGH MILLER.

P.S. — Since writing the above, I have picked up a specimen which, I am pretty sure, you would deem interesting, but for which I have unluckily no room in the box. It contains parts of the tuberculated fossil, and among the rest the teeth of the creature. These last somewhat resemble the teeth of a lobster, being apparently cut out of the solid part of the jaw rather than fixed in it.

H. M.

Miller found more specimens, and more perfect ones, in newly discovered beds of Old Red of Nairnshire, and when Agassiz visited him in 1840, he showed him three well-preserved species of the *Pterichthys*, and the wings of a fourth. To one of these remarkable animals, looking like the letter T, Agassiz has given the appropriate name of *Pterichthys Milleri*. Complete and good specimens were exhibited at the Glasgow Meeting of 1840, and some restorations of the animals were made by

Dinkel, in 1844, for the "Medals of Creation" by Dr. G. Mantell, and were reproduced in the "Vestiges of Creation." But Dinkel, so well trained, and so long Agassiz's artist of fossil fishes, was not successful; and he failed also in trying the restoration of another rather curious form of Old Red fish, the *Coccosteus* or *berry-on-bone*. These two examples show what strange creatures existed during the Devonian period, and the credit of determining their place is due to Agassiz's keen eyes and great knowledge of comparative anatomy; for he did not hesitate, on receiving the first broken and very imperfect specimens, to say that the creatures must have been fishes. As Miller says: "I received new light from the researches of Agassiz which, while it did not show my way more clearly, rendered it at least more interesting by associating with it one of those wonderful truths, stranger than fiction, which rise ever and anon from the profounder depths of science, and whose use, in their connection with the human intellect, seems to be to stimulate the faculties. I have often had occasion to refer to the one-sided condition of tail characteristic of the ichthyolites of the Old Red Sandstone." "It characterizes," says Agassiz, "the fish of all the most ancient formations. At one certain point in the descending scale, Nature entirely alters her plan in the formation of the tail. All the ichthyolites above are fashioned after one particular type — all below after another and different type."[1]

In his preface to "The Old Red Sandstone," Agassiz says: "So true is it that observation alone is a

[1] "The Old Red Sandstone," pp. 115-116, Boston, 1854.

safe guide to the laws of development of organized beings, and that we must be on our guard against all those systems of transformation of species so lightly invented by the imagination." What a prophetic and true sentence against Darwin's "Origin of Species," published fifteen years after. Observations and facts only are given in his "Old Red Fishes," which he has well summarized in the following words: "What I wish to prove here, by a careful discussion of the facts reported in the following pages, is the truth of the law now so clearly demonstrated in the series of vertebrates, that the successive creations have undergone phases of development analogous to those of the embryo in its growth, and similar to the gradations shown by the present creation in the ascending series, which it presents as a whole. One may consider it as henceforth proved that the embryo of the fish during its development, the class of fishes as it at present exists in its numerous families, and the type of fishes in its planetary history, exhibit analogous phases through which one may follow the same creative thought like a guiding thread in the study of the connection between organized being. . . . The facts, taken as a whole, seem to me to show, not only that the fishes of the Old Red constitute an independent fauna, distinct from those of other deposits, but that they also present in their organization the most remarkable analogy with the first phases of embryologic development in the bony fishes of our epoch, and a no less marked parallelism with the lower degrees of certain types of the class as it now exists on the surface of the earth."

The "Monograph of the Fossil Fishes of the Old Red" is more important for the embryologic development, the zoölogical gradation, the geological succession, and the geographical distribution in the past and the present, than the "Origin of Species," by Darwin. It has remained, and will continue to remain, a landmark in zoölogical researches, because nothing in it is left to supposition. Instead of being a work of the imagination, a philosophical dissertation, like the "Origin of Species," it is simply a record of facts and very keen observations; and in science, and more especially in natural history, nothing is of value except exact observations. Agassiz was not an opponent of development; on the contrary, he gave facts in its favour, many years before Darwin did; but he was averse to drawing too hasty conclusions; and he leaned all the time "upon an intellectual coherence, and not upon a material connection"; and he thought that variability seemed controlled by something more than the mechanism of self-adjusting forces. In a word, Agassiz, after his student life, was not a materialist, but a spiritualist, in natural history, an adversary, both of agnosticism and of pietism; for he says: "I dread quite as much the exaggeration of religious fanaticism, borrowing fragments from science, imperfectly, or not at all, understood, and then making use of them to prescribe to scientific men what they are allowed to see or to find in nature" (Louis Agassiz, in a letter to Professor Adam Sedgwick, dated June, 1845 [1]).

[1] "Louis Agassiz," by Mrs. E. C. Agassiz, Vol. I., p. 388.

The "Histoire naturelle des Poissons d'Eau douce de l'Europe Centrale" remained unfinished, and has a rather curious history. Agassiz began it as far back as 1828, when he was a student at Munich, and when his artist friend, Joseph Dinkel, was already making drawings of freshwater fishes for him.

In 1839 appeared the first "livraison" of a folio atlas, published "aux frais de l'auteur," and dedicated to the British Association for the Advancement of Science. This first monograph treated of the salmon family, and was divided into two parts: the first, containing the twenty-seven well-executed and luxuriously printed plates by Dinkel, Sonrel, and Nicolet, illustrating the genera *Salmo* and *Thymalus*, with explanations in French, German, and English, and with a cover designed by Dinkel, representing fishes in all sorts of attitudes and groups, with a boy four years old — the portrait of Alexander Agassiz — fishing on the shore of the Lake of Neuchâtel. The second part of the plates was announced to be issued with the first volume of text; but changes were made, and the text of Vol. I., containing the "Embryologie des Salmones," by C. Vogt, was published in 1842, without plates, the latter being issued in 1848, in Vol. III. of the "Mémoires des Sciences naturelles de Neuchâtel."

Agassiz, under the date of 1845, in the introduction of the "Anatomie des Salmones," by L. Agassiz and C. Vogt, gives the following explanation: "The anatomical studies contained in this memoir were undertaken for the 'Histoire naturelle des Poissons d'Eau douce de l'Europe Centrale, de M. Agassiz,' and were at first destined to form the second volume. Some special circumstances have led the editor to adopt another mode of publication.

"In order to render justice to every one, it is desirable to remark here that the *Osteology* and the *Neurology* are due to the researches of M. Agassiz, while the *Myology*, the *Splanctionology*, the description of the 'sensitive organs' and the *Angiology* have been worked out by M. Vogt. All the plates were drawn by M. Vogt. This work dates as far back as 1843 and 1844, a few observations being added in 1845.

"L. A."

That Agassiz directed the work and freely gave his advice to Vogt, there is no doubt; but in some way Vogt became dissatisfied. He disapproved of the organization and methods of Agassiz's establishment, and was more or less disappointed in his expectations; and, in consequence, in the autumn of 1844, after acting as Agassiz's assistant for five years — during which time he certainly worked most efficiently and very hard — he left him to try his fortune in Paris. Strange to say, the break between Agassiz and Vogt, instead of healing as the years went by, increased to such an extent, that they were very unjust and bitter towards one another. It must be regretted, for nothing really important and seriously affecting either had occurred between them. Agassiz never published anything against Vogt, though Vogt might have shown more discretion in his printed criticism; and I do not hesitate to say that he was unjust and guilty of exaggeration touching some points, and that all that he says about Agassiz's life in America is absolutely erroneous. He disliked Neuchâtel and the Neuchâtelois, and most of his indignation was hurled against them through and to the detriment of Agassiz.

Although Agassiz spent a few days in 1843 on the Aar glacier, his interest in the work going on there was manifestly lessening, and in 1844 he failed to make his usual summer visit.

In July, 1844, Desor, with the permission of Agassiz, published a very interesting and well written volume, entitled, "Excursions et Séjours dans les Glaciers et les hautes Régions des Alpes, de M. Agassiz et de ses Compagnons de Voyage" (Neuchâtel, 12mo). The volume begins with an excellent "Notice sur les Glaciers," by Agassiz, a masterly paper, which gives a scientific turn to the whole work; the rest is written in a picturesque style, and in imitation of the celebrated and popular works of Rudolph Töpffer, the artistic and "spirituel" author of the "Nouvelles Genevoises" and the "Voyages en zig-zag."

It was certainly very generous in Agassiz to allow his secretary to publish at this time all his researches on the glaciers and among the Alps; for it affected the sale of his own "Études sur les Glaciers" (1840) and "Nouvelles Études sur les glaciers actuels" (1847). The last one, more especially, found no sale at all, everything in it having been anticipated by Desor's publication, which, though not so fully developed, rendered Agassiz's work almost superfluous. Desor had taken the lead in the glacial question, and was struggling with its physical problems, for which he was quite as little prepared as Agassiz. Every impartial observer saw this plainly; and it was melancholy to see Agassiz's already straitened resources expended upon almost useless works.

Although the extraordinary meeting of the Geological Society of France, in the Swiss Jura, at Porrentruy, Soleure, and Bienne, in 1838, had much advanced the recognition of the glacial question, it was important that another meeting should be held, this time in the very centre of the phenomena, among the Alps. The city of Chambery, then belonging to the kingdom of Piedmont and Sardinia, was chosen, and the society held its extraordinary session there, during the month of August, 1844. Agassiz presided over several of the meetings; as did also Bishop Rendu. Here these two great masters of the glacial theory met and entirely agreed. After a full and very clear exposition by Bishop Rendu of his "Théorie sur les Glaciers en Général,"[1] the 15th of August, Agassiz says that he agreed "entirely with the theory as it was explained by Bishop Rendu." Numerous adversaries, representing the theories of mud currents, or also of icebergs, tried hard to oppose them; but one after another was silenced by the numerous facts brought forward by Agassiz, Rendu, and others. It was the last strong attempt to resist the glacial theory. Afterward, Élie de Beaumont and his numerous adherents in France and Italy, as well as Leopold von Buch, continued their opposition, in a sort of Platonic

[1] "Bulletin Société géologique de France," 2ième série, Vol. I., pp. 631-636, Paris, 1844.

way, to cover their retreat. But we may truly say that the Chàmbery session of the Geological Society of France was the Waterloo of the mud theory for transportation of boulders.

Agassiz, as usual with him, was very brilliant in his exposition of all the observations he had made on the glacier of the Aar; and Bishop Rendu admirably described the phenomena in Savoy. Agassiz, more especially, insisted that proofs accumulated every year to show that the "Ice-age" extended all over Europe, and that the Alps were formerly a great central mass of ice, extending forty leagues all around, as far as Lyons. Professor Angelo Sismonda, of Turin, continued to maintain that the phenomena did not extend to the southern part of the Alps of Piedmont, until Professor Bartholomeo Gastaldi finally proved beyond question, in 1850, that ancient glaciers occupied the whole valley of the Pô and other valleys in Piedmont, just as they did the valleys of the Rhone, the Arve and the Isère rivers. I well remember those discussions, for I was a hearer of several of them, and can vouch for the splendid part taken by Agassiz in hastening the acceptance of the glacial doctrine. We may say, without any exaggeration, that the interference of Agassiz advanced fully thirty years the recognition of the glacial theory, and that he, and he alone, established the great "Ice-age."

Signs of bad management were visible in more than one direction. The great lithographic institution of Hercule Nicolet was kept running with the greatest difficulty. After bringing about an association of Nicolet with a capitalist, M. Jeanjaquet, Agassiz was constantly obliged to furnish work absolutely unnecessary and very expensive. In a letter to the firm Nicolet and Jeanjaquet, dated 2d July, 1842, Agassiz says: "Vous êtes parfaitement libres de faire ce qu'il vous plaîra à l'égard de vos employés; déjà trop souvent j'ai fait faire des travaux considérables *uniquement* pour occuper vos employés, travaux qui me sont restés des mois et des mois inutilement sur les bras. J'ai fait tirer de fortes éditions d'ouvrages divers, dont je n'ai que peu d'exemplaires placés, pour vous accommoder. . . . J'ai l'honneur de vous prévenir que je désire savoir si je puis compter sur les travaux dont je vous ai parlé lors de ma dernière visite aux Sablons, parce que sans cela j'ai réellement l'intention de les faire faire ailleurs, car je suis sûr d'avance qu'ils me coûteront beaucoup moins." We have here Agassiz's own confession that he undertook some works, solely to give occupation to the too expensive lithographic establishments of the Sablons, — an unbusiness-like proceeding, which was certain to hasten the catastrophe which occurred, in February, 1845, after a struggle of more than a year and a half, when the whole establishment was broken up and disposed of by auction.

Joseph Dinkel, the trusted and true friend of Agassiz, his constant companion since they were students together at Munich, left him to go to England to find work and make a home for himself. He disapproved the leadership of Desor, and foresaw very stormy times for his good friend Agassiz; and he prophesied to an artist friend, who repeated it to me a few years after, that Agassiz would not always submit to such a dictatorship as Desor had assumed, and that it would end in terrible strife. Dinkel clearly saw the game Desor was playing. From the first he did not like him, and it was very painful to him to see Agassiz fall into such hands. He left Neuchâtel, with regret, in the spring of 1844, and many years after acknowledged "that for a long time he felt unhappy at the separation." In most graphic terms he described Agassiz, who, he says, "was a kind, noble-hearted friend; he was very benevolent, and if he had possessed millions of money, he would have spent them upon his researches in science, and have done good to his fellow-creatures as much as possible."[1] Every word is true, and is a noble tribute from one who knew Agassiz most intimately during the time of life when faults of character are most conspicuous, and are easily discovered in the intimacy of friendship.

Still another misfortune befell Agassiz at the close of the year 1844. Gressly, who usually returned to the laboratory in Neuchâtel at the beginning of the winter, did not come back; and it was only after weeks had passed without any tidings of him, that it was learned that the poor fellow was suffering under an attack of religious insanity, and had been placed in an asylum.

[1] "Louis Agassiz," by Mrs. E. C. Agassiz, Vol. I., p. 142.

CHAPTER X.

1845.

"Monographie des Myes," 1842–1845 — The "Nomenclator Zoologicus," 1842–1845 — "Bibliographia Zoologiæ et Geologiæ" — "Iconographie des Coquilles tertiaires réputées Identiques avec les Espèces Vivantes," etc. — The Two Translations of Sowerby's "Mineral Conchology of Great Britain" — Actual Mercantile Value of Agassiz's Publications — Agassiz's Family come to his Help — Great Credit Due to Neuchâtel and its Inhabitants — Agassiz's Last Series of Lectures: "Notice sur la Géographie des Animaux" — Intimate Friendship with Jules Pictet de la Rive — Agassiz's Last Visit to the Aar Glacier — The Meeting of the Helvetic Society of Natural Sciences at Geneva, August, 1845 — A Letter to Pictet, with Biographical Remarks — Biography of Agassiz by Pictet — Agassiz returns all the Specimens borrowed for his Great Palæontological Works.

The year 1845 was spent mainly in finishing the publication of works more or less advanced, and in making a sort of scientific "liquidation," or clearing up. The "Monographie des Myes," a quarto volume, with an atlas of ninety-four well-executed plates, begun in 1842, an excellent and very useful work, containing a number of new, well-defined genera, and which has since been used constantly in conchology, was completed. Alcide d'Orbigny criticised several of the new genera, but Agassiz answered him successfully in his "Introduction," maintaining the value of such genera as *Goniomya*, *Ceromya*, *Arcomya*, *Mactromya*, *Pleuromya*, *Gresslya*, *Cardinia*, etc., and in a letter to Pictet, dated Aug. 15, 1845, he says: "Je crois que vous avez accordé un peu trop d'importance aux critiques que d'Orbigny a faites de quelques uns de mes genres des Myacées. Dans ma 4ième livraison j'ai réfuté ce qui me paraissait exagéré; il y a des remarques justes, mais il y en a quelques unes qui sont complètement erronées."

The manuscript of another work of great importance, of which the first part was issued in 1842, the "Nomenclator Zoologicus," was pushed forward with that strong will which was now and then characteristic of Agassiz. As he says, the work embraces the sources of critical zoölogy: "C'est un travail de patience qui a exigé des recherches bien longues et bien pénibles. J'en avais conçu le plan dès les premières années de mes études et dès lors je n'ai jamais perdu de vue ce projet. J'ose croire que ce sera une digue contre la confusion babylonique qui tend à envahir le domaine de la synonymie en Zoologie" (Letter to M. de Chambrier, President of the State Council of Neuchâtel, April, 1842).

The publication, which is in Latin, is a large quarto, issued in eleven fasciculi; the last one, which treats of the *Coleoptera*, having been published at Soloduri, in 1846; while the "Prefatio indicis universalis" is dated Neocomi, Mense, Decembri, 1845. The "Index" alone comprises 393 quarto pages; a duodecimo edition was also issued at the same time. The general "Prefatio" at the beginning of the work was written entirely by Agassiz. It occupies forty-two pages, rendering justice to all his collaborators, who included Prince Charles Lucien Bonaparte, H. Burmeister, Dumeril, G. R. Gray, Herman von Meyer, Milne-Edwards, Strickland, Charles Des Moulins, etc. This "Prefatio" is dated Neocomi Helvetorum, Febr., 1846, only a few days before Agassiz left Neuchâtel for his journey to America. To give an idea of the great labour expended in carrying the work to completion, it will suffice to say that it contains thirty-one thousand names of genera and families alone, with bibliographical quotations numbering thirty-four thousand titles of works or papers on natural history. In all, the number of quotations is more than one hundred and fifty thousand.

Agassiz had collected for his own private use a catalogue of all known works and detailed memoirs on zoölogy and geology; and, before leaving Europe, he made an arrangement with the "Ray Society," of London, to publish it. Professor H. E. Strickland, the successor of Buckland at the Oxford University, was requested to act as editor; but, unhappily, Strickland was accidentally killed, in September, 1853, while geologizing on the track of the Great Northern Railway, at the mouth of the Clanborough Tunnel, near East Retford, before he had finished the publication of the "Bibliographia Zoologiæ et Geologiæ," based on Agassiz's manuscripts; and Sir William Jardine, the eminent naturalist, and father-in-law of the lamented Strickland, completed the editing of the remaining volumes of the work, which is composed of four octavo volumes, containing the literature of zoölogy and geology until 1846; a most useful publication, dated London, 1848–1854.

In 1845 another memoir on fossil conchology was published by Agassiz, under the title: "Iconographie des coquilles tertiaires réputées identiques avec les espèces vivantes ou dans différents terrains de l'époque tertiaires, accompagnée de la description des espèces nouvelles," in "Nouveaux Mémoires de la Société Helvétique des Sciences Naturelles," Vol. VII. It is, perhaps, the most objectionable paper he ever produced. Starting from a preconceived idea that not a single animal survived a geological epoch, and that no species passed from one formation to another, with his great faculty for differentiating specimens, he easily pointed out a certain number of cases of *Lucina*, *Venus*, *Cytherea*, *Cyprina*, and other acephales, which showed variations, and which, according to his views, demonstrated that the species, instead of being *identical*, were only *analogous*. Deshayes and other conchologists did not accept Agassiz's view, and, in fact, later knowledge has greatly added to the number of species which pass from one formation to another, not only for the tertiary epochs, but also for the Mesozoic and the Paleozoic formations. The complete destruction of faunas and creation of new and entirely different ones, without the survival of a single species, can no longer be defended; more especially in its application to marine animals. As usual, the memoir of Agassiz is beautifully illustrated with fourteen plates, representing with great care all the details of the shells.

During 1845 the last parts of the two translations in French and in German of Sowerby's "Mineral Conchology of Great Britain" were distributed to the few subscribers. The French edition, a large volume with an atlas of 395 coloured plates, is entitled, "Conchologie (*sic*) Minéralogique de la Grande Bretagne," par James Sowerby, "traduction française revue, corrigée et augmentée par L. Agassiz." The German edition entitled, "James Sowerby's Mineral-Conchologie Grossbritanniens, etc., Deutsch bearbeitet von Ed. Desor. Durchgesehen und mit Anmerkungen und Berichtigungen versehen von Dr. L. Agassiz," is also composed of a large volume, with the same atlas of 395 coloured plates. Although the price was considerably lower than that of the original English edition, the two translations did not sell well; especially the French edition, which was and has remained almost absolutely unknown. The undertaking was a great mistake in every way, and both works have remained a drug in the market.

Generally, as years pass, or after the death of an author, some of his publications become rare and valuable, and command a higher price than was asked at the time of their issue. With Agassiz's publications, however, this is not the case; not a single one of his European works is now quoted above, or even at, its price of publication. All are discounted with a fair reduction from the original price, and can be obtained easily of any bookseller in Europe. The same thing has happened also in the case of all his publications in America, with the single exception of his volume on

"Lake Superior," the value of which has risen to two and even three times the price asked by the publisher, when it came out in 1850, and it is now difficult to procure a copy. The explanation of this in the case of some of Agassiz's works, which are really of great scientific value, is, that in their desire to help him, many persons were ready to pay any price asked, and consequently almost all his publications were issued at rather high prices; while others of his publications, although they were expensive, were either not really needed, like the translations of Buckland's and Sowerby's works, or were limited to a too small circle of naturalists to secure a large sale.

Better management would have prevented Agassiz from running into debt on account of his numerous publications. At the same time that he was issuing his works with such losses, works of the same sort were published in France, not only without loss, but even with profit from the very beginning of the undertaking. I refer to the great work of Deshayes, "Description des Coquilles fossiles des Environs de Paris," and more especially the "Paléontologie française," by Alcide d'Orbigny. Agassiz differed from them, also, in his method of working, and in his domestic arrangements, for both Deshayes and d'Orbigny worked alone, without assistants of any sort, except their artists; and their establishments at Paris were extremely modest, and limited to what was strictly necessary.

In 1845 the pecuniary position of Agassiz became very serious, and his family were obliged to come to his assistance, which they did with great generosity. All his numerous and bulky publications were put into the hands of the rich firm of Jent and Gassmann, booksellers and publishers at Solothurn; securities were given to his creditors, and everything was most honourably arranged to relieve him from his immediate distressing position.

If, however, his Neuchâtel establishment was a failure pecuniarily, scientifically it was a success unique in natural history. The result of his fourteen years' residence at Neuchâtel was the publication of more than twenty volumes, with two thousand folio or octavo plates, and many separate papers; all were well written, beautifully printed, and profusely illustrated with most exact drawings — a record so creditable that it gave a just celebrity, not only to Agassiz, but also to Neuchâtel, at that time a small town of less than six thousand inhabitants. The "Neuchâtelois" may well be proud of such a performance; their great liberality toward science, and their appreciation of the rare value of Agassiz, made it possible for him to prosecute with unimpaired vigour his remarkable scientific researches famed the world over.

That Agassiz thought that he was acting wisely in receiving Vogt and Desor at his table as regular boarders, and giving a room in his apartment to Desor, there is no doubt. But, in the long run, the scheme proved expensive, and most harassing to his wife Little by little, the characters of both Vogt and Desor came out; jokes of doubtful politeness were indulged in; remarks rather satirical, cynical, and anti-religious were not rare. Vogt, more especially, never missed an opportunity to make a "bon mot" at the expense of the

Bible, turning into ridicule all religious beliefs and practices. Mrs. Agassiz, being a religious woman and bred in a totally different atmosphere in her own home at Carlsruhe, was very sensitive to these sarcasms. Finally expenses and difficulties reached such a climax that a crisis became inevitable. Mrs. Agassiz's health was poor; and the announcement, by newspaper, all over Germany, of a royal gift by the king of Prussia to allow Agassiz to make a journey to America, was hailed as a proper moment to join her own family at Carlsruhe.

In a letter, dated Carlsruhe, 16th of March, 1845, her devoted brother, Alexander Braun, wrote that all was ready at his home to receive her. (Alexander Braun's "Leben," pp. 378, 379.) Taking her children with her, first on a visit to the excellent mother of Agassiz at Cudrefin, at the old Dr. Mayor's house, Mrs. Agassiz then left Neuchâtel early in May. It is the most painful incident in the life of the great naturalist. That misunderstandings and difficulties developed mainly by extravagance in the interest of natural history should have had such a final result, is most pitiable and to be regretted. These explanations are not meant to excuse the faults committed by Agassiz at this time of his life; they show, however, how he fell into errors, and how he might easily have avoided them. They have been rendered necessary by what has been said, rather bitterly, by the biographer of Desor ("Edward Desor: Lebensbild eines Naturforschers," von Karl Vogt, Breslau, 1882).

In the spring of 1845, Agassiz delivered his last public course of twelve lectures on the " Plan de la Création," showing the successive development of organized beings. It was followed with more attention and by a more numerous audience than any of his previous annual series of lectures. The news that he was to undertake a journey to the New World under the auspices, and with the help of the king of Prussia and prince of Neuchâtel, who contributed from his private purse three thousand dollars, caused a surprise mingled with fear that he would probably never return to resume his position at Neuchâtel. Everybody in Neuchâtel highly appreciated, not only the great savant who was truly the founder of the Academy, — which, but for him, would not have been established for years, — but also the friend and charmer so highly esteemed and beloved, and went anxiously to hear him once more; anticipating, with good reason, that this last course might be regarded as his scientific testament.

Agassiz took care to dictate his last lecture, and published it in the first number of the " Revue Suisse," just transferred from Lausanne to Neuchâtel, in August, 1845. The title of the lecture is: " Notice sur la Géographie des Animaux, par L. Agassiz"; and it begins with the following sentence: " All organized beings, plants as well as animals, are confined to a special area [or, as he calls it, " ont une patrie"]. Man alone is spread over the whole surface of the earth." Strange to say, one of his first impressions, after studying the different races of man he met with in America, led him to reverse this opinion, and a few years later he published his remarkable and scientifically very frank paper, "Sketch of the Natural Provinces of the Animal World and their Relation to the Different Types of Man," Cambridge, 1853. Agassiz was too good a naturalist, too much accustomed to differentiate animals, to accept unity in the genus *Homo*, and when converted to the views of Dr. Samuel Morton of Philadelphia on the different types and diversity of man, he frankly proclaimed his change of opinion to the scientific world, with the same earnestness with which eight years previously he maintained the old creed of a unique species; and when, a few years later, he heard of the discoveries of the fossil man of the Quaternary epoch, he accepted it at once, delighted to learn that a man was in existence and saw the great glaciers, of which he was the first to conceive the existence before the present epoch.

I may add a personal reminiscence of the first time I saw Agassiz, when I presented to him a letter of introduction from his friend Jules Thurmann. He had close by him on his desk a pile of copies of this notice on the geography of animals, and taking one, he wrote my name on the cover, and offered it to me. I have ever kept that first gift of Agassiz — followed by many others, for he always from that time gave me all his publications — as a souvenir of one of the most fascinating men I have met in my life; for such was the impression he made on me; an impression which has remained unimpaired, and indeed constantly deepened, until the last day of his life.

During 1845, the friendship which had existed for at

least twelve years with Jules Pictet de la Rive of Geneva, became intimacy, and remained so until the end. These two savants had many similar qualities, and it is not surprising that when they met they became the best of friends. For Pictet life was never difficult. Son of an old and wealthy family, he married early the granddaughter of the celebrated Madame Necker-de-Saussure, and became one of the richest men of Geneva; while, on the contrary, Agassiz had to struggle all his life against poverty. However, both of them spent largely for science, and were never so happy as when they were able to secure at any price rare and well-preserved natural history specimens. Although rich, Pictet always worked very hard, being second only to Agassiz in this respect, and not far behind him. He conceived and published the first manual of palæontology in four volumes; and the first copy was sent to Agassiz, who wrote at once a review of it for the "Bibliotheque Universelle." The following is a letter from Agassiz on the subject:—

NEUCHÂTEL, 7 Mai, 1845.

F. J. PICTET,
Geneve.

Mon cher ami,—C'est à vous plutot qu'a Monsieur de la Rive que j'adresse l'analyse que je viens de faire de votre ouvrage ("Traité élémentaire de Paléontologie"). Vous voyez que j'ai tenu parole de ne prendre que le temps matériellement nécessaire à sa lecture pour la rediger. Aussi ma notice doit se ressentir de cette precipitation, et c'est ce qui m'a fait décider de vous la soumettre d'abord. Corrigez et changez ce que vous voudrez, je n'ai pas l'esprit assez repose pour avoir pu faire quelque chose de complet, quoique j'aie lu votre beau livre bien attentivement. Je ne doute pas que cet ouvrage n'ait un grand succès, et je vous engagerais fort à en provoquer la traduction en anglais et en allemand.

J'aurais eu un chapitre à écrire sur la nature des progrès organiques, réalisés dans la série des terrains, mais cela m'aurait entrainé à m'écarter trop de votre ouvrage, et c'est ce que je n'ai pas voulu faire.

Vous seriez bien aimable de venir me voir bientôt; quant à moi je doute de pouvoir aller encore une fois à Genève; je suis accablé de travail. Mais proposez cette course à Favre et venez bientôt; ce serait un grand bonheur pour moi de vous revoir avant mon départ qui est fixé à la fin de Juin. [He did not leave, however, until ten months later.]

Je vous enverrai sous peu un long mémoire sur la question des Coquilles Tertiaires réputées identiques avec les vivantes, que je viens de faire imprimer.

Mille amitiés à M. de la Rive. J'ai d'excellentes nouvelles de Vogt, qui travaille comme un forcené à Paris, je lui ferai parvenir l'exemplaire du Traité (de Paléontologie) que vous lui destinez et qu'il n'a sûrement pas davantage que moi.

Votre tout dévoué,

LS. AGASSIZ.

Dites moi si vous me pardonnez mes critiques.

A last visit to the glacier of the Aar, at the beginning of August, 1845, was made in order to transfer all the observations to Daniel Dollfus-Ausset, who had generously offered to continue them at his own expense, and who did so with great perseverance, during sixteen years, until 1861. (See "Glaciers en activité," in "Matériaux pour l'Étude des Glaciers," par Dollfus-Ausset, Vol. V., Paris, 1870.)

As soon as Agassiz returned to Neuchâtel, he again left to attend the annual meeting of the Helvetian Society of Natural Sciences, held at Geneva the 11th, 12th, and 13th of August, 1845. There, although only a middle-aged man, he seemed like the leader of the meeting. He spoke first, at the general session, on the structure of the fins of fishes; and then, at the special sections of physics, he gave an account of his researches during the last three years on the glacier of the Aar, dealing more especially with the motion of the glacier, its structure, the ablation of the surface, the meteorology, etc. Discussion followed, in which Jean de Charpentier, the founder of the glacial theory, and Venetz, son of the first promoter and discoverer of the existence of ancient and immense glaciers in the Rhone valley, took part and gave new proofs of the great value now attached to their first observations. Leopold von Buch, present at the meeting, did not approve all that he heard respecting glaciers, and left, rather indignant at the evidence of the great progress made; for at this time, all the Geneva naturalists, with the exception of Jean André de Luc, then an octogenarian, were converted to the new theory. Arriving at Zurich a few days after the meeting was over, von Buch called on Arnold Escher von der Linth, as he was accustomed to do almost annually, and begged Escher to take him on an excursion among the Alps of the Primitive Cantons of Switzerland, making the one condition, however, that Escher would not once speak of anything relating to glaciers and glacial action. Escher, who respected and loved von Buch, as the best friend of his deceased father, promised, and kept his word, notwithstanding that he was himself one of the best and first-converted glacialists, and that at every step he found most undeniable proofs of the great extension of glaciers. A few years later, Escher told me that it required the greatest self-control he was ever able to exercise; and that nothing could induce him to attempt it again.

But to return to the Geneva meeting: Agassiz made a third communication on the brain of fishes, and noted the existence of a very large pot-hole ("Marmite des Géants") above the Handeck Fall in the Bernese Alps. At the great dinner, given to the Helvetian Society at the "Hôtel de la Navigation," Agassiz was toasted, with the following remarks: "To the learned and amiable professor of Neuchâtel, M. Agassiz, who is on the point of undertaking a far-distant journey, where our sympathies will follow him," etc.

The impression made on Agassiz was very strong, as shown by the letter he immediately wrote to Pictet:—

NEUCHÂTEL, le 16 Août, 1845.

JULES PICTET,
Genève.

Mon cher ami,—Je ne veux pas tarder à vous envoyer Miller ("The Old Red Sandstone"), afin que vous le receviez avant votre départ pour Naples. Renvoyez moi le dès que vous le pourrez.

Je suis rentré chez moi hier, comme je me l'étais proposé, et j'ai trouvé ma mère qui m'attendait déjà depuis la veille; vous voyez que c'eut été très mal de ma part de prolonger mon séjour à Genève. Il fallait vraiment un pareil motif pour me donner la force de me séparer de vous. Cette réunion a laissé dans mon cœur des souvenirs ineffaçables; veuillez répéter encore à tous nos amis communs, et en particulier à M. de la Rive, à Favre, à M. Marcet, combien j'ai été touché de leur accueil amical et de toutes les marques d'amitié qu'ils m'ont données.

J'ai déjà parcouru quelques feuilles de votre troisième volume (*Traité de Paléontologie*); plus j'apprends à connaître votre livre, et plus je suis convaincu qu'il aura un grand succès. . . . A propos, j'ai oublié de vous demander si vous avez donné suite à la demande qui vous a été faite de rédiger une notice biographique sur mon compte, dans la *Revue Romande* ou je ne sais quel autre recueil. Si cela est, dites-moi où je la trouverai.

Votre tout dévoué,

LS. AGASSIZ.

The biography by Pictet, with an excellent portrait of Agassiz, was communicated in manuscript to Agassiz, who, in returning it, wrote in the following terms: —

NEUCHÂTEL, 25 Août, 1845.

JULES PICTET,
Genève.

REMARQUES:

Verso de page 1. — "Cette correction, etc."; à effacer, on pourrait croire que mon père m'avait à demi assommé,[1] et personne n'était moins sévère que lui.

Page 2. — "d'eau douce"; ajoutez: pour lequel il recueillit d'importants matériaux dans le Rhin et dans le Necker, qu'il put comparer plus tard avec ceux du Danube et de l'Isère, pendant son séjour à Munich et à Vienne.

Verso de page 2. — Ajoutez au bas de la page: Ce goût pour l'observation fut encore augmenté par les nombreux voyages qu'il fit dans le midi de l'Allemagne, et en particulier dans les Alpes du Tyrol, où il se familiarisa avec l'étude des plantes, sous la direction d'un de ses condisciples, M. Alex. Braun, devenu depuis botaniste distingué. Ces connaissances lui furent plus tard d'une immense utilité pour l'étude des plantes fossiles.

Page 4. — "Vogt." La large part que j'ai faite à Vogt dans la publication de l'embryologie des Salmones, que nous avons poursuivie pendant toute une saison en commun, et que je l'ai chargé plus tard de terminer, tandis que j'aurais été pleinement en droit d'après

[1] Here is the corrected sentence: "Il (Agassiz) raconte lui-même que l'amour de la pêche l'entraînait quelquefois trop loin, et que la seule punition qu'il reçut jamais de son père lui fut infligé parce qu'il s'était imprudemment embarqué dans un petit bateau pour la pêche du brochet."

ces antécédents de les publier sous nos noms réunis, ne doit pas être un motif pour m'exclure de toute participation à ce travail; ainsi modifiez légèrement cet article, d'après la préface de l'ouvrage rédigée par Vogt lui-même.

Page 9. — Vous citez ici pour la deuxième fois l'ouvrage de Charpentier et vous ne mentionnez pas même le mien qui était déjà, lorsqu'il parut en 1840, appuyé sur plus de faits que ceux que décrit Charpentier. J'y donne déjà des chiffres, et les premières planches que l'on ait possédées sur les glaciers, faites en vue de faire connaître leur structure, la disposition des moraines, l'action des glaciers sur le sol, les roches polies, etc., etc.

Ce fut déjà en 1840 que j'allai visiter l'Ecosse pour y chercher des traces de glaciers et que je démontrai leur présence dans une foule de ces belles vallées, tant d'après l'arrangement des moraines qui les traversent, que d'après la nature des polis de leurs parois rocheuses et des galets de leur fond. Je les observai aussi en Irlande, en Angleterre dans la région des lacs, et plus tard dans la Forêt Noire. Et c'est à ces observations qu'est dû l'intérêt général qu'a pris la question des glaciers.

Dernière page. — Ajoutez : Sous les auspices et aux frais du Roi de Prusse, auquel j ai dû de flatteuses distinctions, ou quelque chose d'analogue ; ce sera utile pour l'avenir.[1]

Voilà bien des observations, mon cher ami, mais vous les introduirez encore plus brièvement dans votre notice qui me fait grand plaisir, et pour laquelle je vous remercie vivement. Si j'ai fait une petite note pour les *Études sur les glaciers*, c'est que j'ai le sentiment qu'à l'égard de ce livre, on n'a pas été juste à mon égard et que de toutes parts on lui a jeté la pierre contre ; les uns disent que j'en avais emprunté le contenu à M. de Charpentier que je n'avais pas pourtant visité qu'en 1836, tandis que mon ouvrage est de 1840, et est le résultat de mes courses et de mes observations propres ; de là la divergence sur tant de points avec de Charpentier qui n'a publié qu'un an plus tard ; les autres, et Forbes en particulier, même en 1842, m'ont refusé la connaissance de tout fait qui n'était pas

[1] It is evident that Agassiz, at that time, still hoped to be called to a professorship in the University of Berlin.

mentionné dans mes *Études* et cela même pour des faits que je leur ai montrés le premier. Ce n'est donc pas pour une misérable gloriole que je réclame, mais par un sentiment de justice.

Demain ou après-demain je vous enverrai mes *Myes*.

Adieu, mon cher ami.

Tout à vous,

Ls. AGASSIZ.

P.S.—Voudriez-vous dire à Favre ou à son frère, si Alphonse est aux Diablerets, qu'alors même que je ne réponds pas immédiatement à sa lettre, il peut compter que je lui donnerai les renseignements qu'il me demande pour son voyage dans le Nord, avant mon départ.

Pourrais-je obtenir de l'éditeur de votre notice d'en avoir quelques exemplaires pour distribuer à mes amis?

Adieu, bon voyage si vous allez à Naples, mes amitiés au Prince de Canino.

The biography of Agassiz by Pictet is almost unknown, on account of its publication in an album, which had a very limited circulation, confined to French Switzerland, and among a circle of subscribers residing in villas round the shore of the Lake of Geneva. Of the many biographical notices, published either during the life or after the death of Agassiz, it is by far the best; and I cannot do better than to quote the first and last sentences,—two admirable pages of true and just homage to the great naturalist. Pictet says:—

Parmi les savants dont la Suisse a pu avec raison s'honorer dans ces dernières années, Agassiz est certainement un de ceux dont la réputation est la plus populaire. Des travaux scientifiques remarquables, empreints de ce mélange d'imagination et de jugement qui caractérise les créations brillantes et durables, une grande persévérance dans l'étude des faits, une éloquence chaleureuse et entraînante, justifient amplement cette réputation, à laquelle ses études sur les glaciers, plus à la portée de tout le monde que ses autres travaux, ont ajouté un nouvel éclat. Qui, en effet, parmi les amis des Alpes, ne s'est pas intéressé à cette petite réunion d'hommes, liés par l'amour de la science, transportant leur laboratoire dans ces Hautes Regions Glacées, décrivant en artistes les beautés spéciales de ces vastes solitudes, y exercant une large hospitalité et déployant dans leurs travaux une persévérance, une ardeur et quelquefois une hardiesse bien faite pour captiver l'attention des plus indifférents? . . . Nous avons cherché à esquisser ici la vie déjà si remplie de notre savant compatriote, nous aurions voulu oser pénétrer encore plus avant, et raconter à ceux qui ne le connaissent pas, son caractère aimable et attachant, son ardeur dans tout ce qu'il entreprend, sa vivacité dans la discussion unie à la politesse du cœur et en un mot toutes les qualités qui lui ont créé partout des amis et qui l'ont fait l'âme de réunions des naturalistes suisses qu'il vivifie par sa présence.

Such appreciation, coming from so independent and just a naturalist as Jules Pictet de la Rive, shows what a strong hold Agassiz had upon his countrymen, when hardly in middle life; indeed, before he was forty years old. Every one in Switzerland felt that so small and modest a place as Neuchâtel could not retain him any longer. Even Swiss naturalists saw plainly that Switzerland was too small a field for the indefatigable activity of a man so gifted, and that his proper place was either at Paris, or in a new and great country like the United States of America; and when he left, everybody knew that it was a final departure, and that Agassiz was lost to the fatherland.

Agassiz's letter to Pictet is also most important, because it gives an inside view of several occurrences, more especially of the difficulty with de Charpentier. Agassiz did not realize the impression made on many by the impropriety of his publishing before de Charpentier a volume of "Études des Glaciers." Many of Agassiz's best friends regretted it sincerely, and there is no doubt that it was a mistake on his part not to have waited until de Charpentier had issued his volume.

A last duty remained to perform before saying good by to Neuchâtel. It was to return all the specimens of fossils so generously lent by public establishments and private individuals for his palæontological works. It was not a small undertaking; for Agassiz, with his eagerness to collect all the material he wanted, asked for and collected around him a quantity of specimens which he was unable to make use of. However, everything was carefully packed and sent safely to its destination with the thanks of the professor. And since that time all such specimens are quoted in both public or private collections as determined by Agassiz; every one being justly proud to have helped the author of the "Poissons fossiles," of the "Echinodermes," and of the "Myes."

CHAPTER XI.

1846.

Departure from Neuchâtel, March, 1846 — Arrival in Paris and Sojourn at the "Hôtel du Jardin du Roi" — "Nouvelles Études sur les Glaciers actuels" — The Glacial Theory before the Geological Society of France, at the Meeting of the 6th of April, 1846 — Agassiz's "Catalogue Raisonné des Echinodermes" — His Work in the "Galerie de Zoologie" and among the Private Collections of Brongniart, de France, Deshayes, d'Orbigny, de Verneuil, etc. — Desor's Presumption, in putting his Name on the Title Page, without Agassiz's Knowledge — Attentions paid to Agassiz by Thiers — Indirect Offer of Official Positions at Paris declined — Short Visit to England, to meet Charles Lyell — On a Cunard Steamship from Liverpool to Boston.

At the beginning of March, 1846, Agassiz left Neuchâtel, never to return, except for a single very short visit, of a few days, in 1859. At two o'clock in the morning, he took the stage for Bâle, en route for Carlsruhe; and, notwithstanding the early hour, the post-yard was filled with his many friends, colleagues, and students, the last coming in a body, with torch-lights, and giving him a parting serenade. Although he spoke of his return, and of resuming his scientific work at Neuchâtel, every one felt that the departure was momentous, and that Neuchâtel was losing the man who had given it a world-wide reputation as a centre of science, never before equalled in Switzerland, in connection with such a small town, such a limited academy, and in so short a time.

In order to understand what follows, it is necessary to say a few words concerning the material difficulties under which Agassiz laboured during his fourteen years' residence at Neuchâtel. His very small salary, of eighty louis (Neuchâtel currency), and a few years later of one hundred and sixty louis, was hardly sufficient to defray his household expenses, even if he had limited them strictly to his family. But very soon he largely increased all his expenditure, both for his publications and for his assistants. At first, his sister and wife helped him, and his friend Louis de Coulon assisted him in bibliographic work, and in collecting under his direction. But when he became interested in glaciers and the glacial question, it was too great a task for his voluntary assistants, and, in addition, new duties obliged Mrs. Agassiz to give up drawing and writing for her husband. If Agassiz was an indefatigable worker, when busied in the observation of new facts, he was too impatient, and always carried too far by new schemes, to write books, or even memoirs. As he himself says, it was very difficult for him to sit down at his desk and write all he had observed and knew on a subject. "Je ne suis pas un cul de plomb comme Richard Owen," the great English palæontologist. He always envied this faculty, so strongly characteristic of Professor Owen. But it was vain for him to try to acquire it, for he soon fretted, was extremely nervous, and finally left the work to others to finish, or abandoned it altogether, never returning to it. With such a disposition, Agassiz was much in need of a secretary and assistants able to understand his instructions and to carry out and finish his numerous schemes. He successively added several assistants. In fact, the apartment of the professor was a sort of "Pension bourgeoise" for naturalists and artists; for, besides the regular inmates, there was a constant arrival of friends, and of members of the Agassiz family, who were quite numerous around Neuchâtel, and of foreign naturalists, such as the two Schimpers and the two Braun brothers. Of course, the one hundred and sixty louis of his salary were soon exhausted in keeping such an establishment, and needed additions of money were lacking all the time. Agassiz very quickly expended his share of his inheritance from his father, and then all his family were obliged to help him; which they did at first with pleasure, and afterward with some reluctance.

The Neuchâtel burgesses, and more especially all the wealthy families, who had contributed to the subscription for founding his professorship of natural history, were ready to help him, and very generously contributed money for each new scheme brought before them by Agassiz. But as soon as one scheme was fairly started, another, absolutely unexpected, was added to the burden. And, as one of the most liberal of those naturalists of Neuchâtel says, "We were ready to help Agassiz with money; but there was no end to his constant needs. He had already expended, in advance, all we were glad to offer him, et c'était toujours à recommencer." In fact, Agassiz had exhausted all his credit, when he left Neuchâtel, having made use, one after another, of each of his friends, and of his whole family. And all for science! for he had few needs, and was by no means extravagant in personal expenses. Always generous when he had money in his hands, he distributed it to his assistants, draughtsmen, and lithographers, never thinking of himself and of his own family, until all others had been supplied. On the whole, Agassiz was a very rare character, — always hopeful, but a great dreamer; and he acted, all his life, as if he knew with certainty that a great lump of gold belonging to him was lying somewhere behind an enormous boulder, and that he had only to extend his hand behind the boulder, and fill his pockets with as much as he wanted. And, curiously enough, this dream of his was fully realized, only it was at the end of his life, and for the benefit of his children. And so was fulfilled Humboldt's prediction, in a letter dated Berlin, June 17, 1838, that "he was certain that there was gold somewhere in his *polished rocks.* I should like to find the secret which you possess, to work all those mines." For it is under, and even in, *polished rocks* of the great North American glacier extending from Greenland to Minnesota that Agassiz's great gold lump lay.

When Agassiz left Neuchâtel, it was arranged that Desor and Girard should pack up about two hundred volumes, — the most necessary works for reference on glaciers and fossil echinoderms, — and leave all the rest of Agassiz's already large library in charge of

William Huber, the librarian, with directions to continue the bibliographic collection of titles for Agassiz's great manuscript list, forming his "Bibliographia Zoologiæ et Geologiæ," and then hasten to Paris to meet Agassiz on his arrival there.

The son, Alexander, then a boy of eleven years, was left at Neuchâtel, to pursue his studies at the College. The two daughters and Mrs. Agassiz were already living at Carlsruhe with Alexander Braun, the always trusted friend of Agassiz and the excellent brother of his wife. Having disposed as satisfactorily as possible of all his affairs and the numerous persons more or less dependent on him, Agassiz took his departure, with a heavy heart and great anxiety as to his future. He knew too well that it was impossible for him to return and assume again the same position, — a position inadequate to his wants and his aspirations as a savant and as the head of a family. The world was open before him, to be sure; but all was uncertainty. However, his will was strong to conquer a position; and with that determination constantly in view, he began life again at the ripe age of thirty-nine years.

After a few days passed with his family at Carlsruhe, Agassiz arrived in Paris at the end of March, staying, as he was accustomed to do, at the old "Hôtel du Jardin du Roi," rue Copau (now rue Lacépède), near the Jardin des Plantes. There he was received by Desor and Girard, to whom were added Karl Vogt, at that time a resident of the hotel, and Dickmann, one of Agassiz's artists.

At once Agassiz started several works; first, an

octavo volume on the glaciers, and second, a "Catalogue raisonné des Echinodermes vivants et fossils." After his publication of "Études sur les Glaciers" (1840), Agassiz began in 1841 a new series of researches and observations on the structure of ice, the temperature, the annual progression, and the daily movement of glaciers; and it was the result of these four years of constant study on the glacier of the Aar that he wished to present to the scientific world.

A well-known Paris publisher, M. Victor Masson, purchased Agassiz's manuscript, the first fruit of his arduous toil that Agassiz had succeeded in thus disposing of; but, unhappily, the transaction proved an unfortunate one for the publisher, who lost heavily, the failure being due partly to political trouble in France in 1848, a short time after the work was issued, partly to its incompleteness. According to the announcement, it was to be composed of three parts, of which the first only was published; the contemplated second part was to be furnished by Arnold Guyot, on the distribution of boulders round the Alps, and the third part, on the geographical distribution of old glaciers all the world over, by E. Desor. Guyot and Desor contented themselves with a few short papers, published in the "Bulletin de la Société des Sciences Naturelles de Neuchâtel," 1847, on the erratic boulders of the basins of the Rhone, Rhine, and the Pennine Alps; and in the "Bulletin de la Société Géologique de France," on the glacial deposits of Scandinavia, and the erratic or Quaternary of North America.

As usual, Desor wrote the first part, under Agassiz's

direction and supervision. Chapter by chapter, Agassiz looked over the manuscript, correcting with pencil, and indicating additions to be made. The manuscript was finished before Agassiz left Paris, and went to the printer between November, 1846, and April, 1847; first under the direction of Desor, who left Paris at the end of February, 1847, and afterward under the direction of Charles Martins, who wrote the introduction and finished the excellent list of works on the present glaciers. Thus the volume is a rather composite one, through the collaboration of Desor and Martins, and as a whole, is less important than Agassiz's first volume on the glaciers, although it contains many new facts. The truth is, that Agassiz and Desor were not physicists; and although Martins and Bravais, who were good physicists, helped them with their advice at the glacier of the Aar, they failed to recognize the plasticity of glaciers, as Bishop Rendu and James Forbes had done in the case of the Savoy glaciers; and it was reserved for the great English physicist, John Tyndall, to solve the problem of the conversion of snow into ice by pressure, to find the cause of glacier motion in pressure, regelation, crystallization, and internal liquefaction, — a splendid discovery which was made between 1856 and 1859, and published in 1860, in a work entitled "Glaciers of the Alps."

Beside the publication of the volume on the glaciers, Agassiz, during his stay in Paris, greatly advanced the acceptance of the glacial doctrine by all unprejudiced geologists. In a communication made before the Geological Society of France, at the meeting of the 6th of

April, 1846, he discussed, with more care, if possible, than usual, all the plain facts observed on the present glaciers, as regards polishing of rocks, directions of striated marks, "cailloux striés, boue glaciaire," transportation of boulders, etc. For we must keep in mind that everything was contested and often denied by the opponents of the glacial theory. Agassiz had before him, however, an audience suited to his wishes. De Beaumont, the great adversary of glaciers, was there; also de Beaumont's collaborator and right arm, Dufrenoy, besides some partisans of his own views, among them Constant Prévost, Deshayes, Martins, Bravais, Dollfus-Ausset, d'Omalius d'Halloy, and Major Leblanc. It was a very important meeting, for Agassiz was able to answer every objection. De Beaumont, who was always very cunning when in the presence of original and able observers, preserved a discreet silence, and let all the heat of the discussion rest on Dufrenoy, contenting himself with smiling and nodding his approval. It was a curious duel. Dufrenoy, always sceptical, but amiable, and rather inclined to be humorous, asked if the "cailloux striés" were truly a good indication of the existence of old glaciers. "Yes!" was the answer. "They are the characteristic fossils of a glacier." Little by little, the audience of eighty persons, all good geologists, came round to Agassiz's views. It was a marked success; so much so, that de Beaumont left the room before the end of the meeting; and Dufrenoy, when the meeting was over, said aloud to Agassiz, referring to his collaboration and companionship with de Beaumont during twenty-five years,

"Croyez-vous que j'ai été toujours à la noce avec lui;" showing how much he had to endure from the disposition of his colleague in the construction of the Geological Map of France.

On this day the glacial theory at last gained the ascendency in France. Dé Beaumont, for two years longer, continued an underhanded opposition by means of some of his favourite pupils, Messrs. Durocher and Frappoli. But Charles Martins, a remarkable speaker and good writer, took the question where Agassiz left it, and easily extinguished all opposition. Now it may seem strange to many that such a clear question, with such admirable and visible proofs, should have encountered such a powerful opposition, and arraigned against it such geologists as Alexander von Humboldt, von Buch, Élie de Beaumont, and Murchison. Geology is too vast for any one man, whatever his intellectual capacity and knowledge, to be a good judge and an expert on all the questions which arise. At the beginning of the création of modern geology it was the custom for every one to give his opinions on each point. In this way, a number of errors were accepted as facts; and it required generations of able observers to remove these great obstacles to the progress of geology. The belief in the transportation of boulders by great mud currents, in connection with the universal deluge of the Mosaic tradition, was so deeply implanted in the minds, even of savants, that it was not an easy task for Venetz, de Charpentier, and Agassiz to yproot it. It laid upon them a quarter of a century of hard work and harder fighting.

From the time of his first establishment at Neuchâtel, Agassiz had taken great interest in the echinoderms, publishing, from 1833 to 1845, numerous and most important memoirs on the subject. His stay at Paris was an opportunity long looked for, and he seized upon it with his usual enthusiasm. All the public and private collections were at his complete disposal. The Jardin des Plantes, with its vast wealth, known and unknown, was thrown open to him. The old gallery of zoology, just opposite the "Pitié Hospital" had its best room barricaded; and drawers filled with specimens, barrels of all shapes, containing collections of marine animals from all parts of the world, and never opened until now, were brought from cellars and garrets, and arranged in front of the usual collection of echinoderms exhibited to the public. Agassiz placed the specimens on long tables; and there, with the help of his friend Valenciennes, professor of conchology, and his assistant, Louis Rousseau, — a brother of the great landscape painter, Theodore Rousseau, — he began classification and determination, dictating to his secretary, Desor, the descriptions of families, genera, and species. Sometimes his enthusiasm was raised to perfect rapture, when some new species or a new genus was found in one of the barrels brought up from the Pacific Ocean by exploring expeditions of the end of the last or the beginning of this century. It was interesting and also amusing to see him with a sea-urchin in one hand, and a lens in the other, analyzing each organ and each part of the animal, with that accuracy of description for which he was justly celebrated;

and after looking at the label of the barrel, he would sometimes exclaim, "Why! it was collected by Quoy and Gaimard, or by Humbron, or some one else, on the shore of Tasmania or New Zealand, during the voyage round the world of de Freycinet, or Duperé, or Dumont d'Urville," etc. Happiness beamed on his face; and satisfaction was seen in every movement, exclamation, and posture. What an admirer of natural history objects! It was impossible to resist feeling interest in his work. He excited the curiosity of every one in the gallery, and even the guardians and porters were deeply affected and attracted around the professor. The guardian, or janitor, named Philippe Pothau, so well known by all zoologists who have studied, or even only passed through the collections of the Jardin des Plantes, was in ecstasy and rapture before Agassiz. He was not accustomed to see such enthusiasm, Valenciennes being the most prosaic and unmovable of men, and all the other professors of the Jardin des Plantes being either very sceptical, or too busy to pay much attention to the treasures under their guardianship.

The private collections at Paris were then more numerous and more important than at the present time. The impulse given to the study of palæontology and geology by Cuvier and his school had not yet died out. His principal collaborator Alexandre Brongniart was still alive; and on two successive Sundays he himself exhibited to Agassiz his fine collection of fossil echinoderms, some of which were the types described by Lamarck and himself in his celebrated "Géologie des Environs de Paris." Defrance, one of the ablest and most modest of all French palæontologists of the first half of the nineteenth century, was also still alive, and with his printed list of fossil remains entitled, "Tableau des corps organisés fossiles," etc., in his hands, he pointed out each of the echinoderms to Agassiz. Besides those two collections, so important on account of the types they contained, Agassiz studied, one after another, the fine collections of Alcide d'Orbigny, Deshayes, Michelin, Graves, de Verneuil, d'Archiac, as well as the public collections of the École des Mines, la Sorbonne, and the École Normale. It was a rare enjoyment for Agassiz.

He himself wrote, without any aid from his secretary, the "Résumé d'un travail d'ensemble sur l'organisation, la classification et le développement progressif des Echinodermes dans la Série des terrains"; a masterly review of his knowledge of the Echinidæ, and read it before the Academy of Science of the Institute, of which he had been a corresponding member since April, 1839. Printed first in the "Comptes-rendus de l'Académie," Vol. XXIII., it was reprinted with very few alterations and addition in the "Annales des Sciences naturelles," as an introduction to the "Catalogue raisonné des familles, des genres et des espèces de la classe des échinodermes, par MM. L. Agassiz et E. Desor." The secretary and assistant of a savant has no scientific right to authorship in the publications made by the savant, though generally the savant says in the introduction, or in the body of the work, that he has been helped by his assistant. Agassiz refers several times in the introduction of the "Catalogue raisonné des Echinides," to Desor and his help; and it was the only recognition really due. But Desor, without asking permission, took upon himself to add his name, as one of the two authors of the "Catalogue," a high-handed proceeding which did not come to the knowledge of Agassiz, until May, 1848, when he received the fifty separate copies printed for his private use. It is not surprising that Agassiz resented the presumption and expressed his disapproval in his great work: "Contributions to the Natural History of the United States of America," Vol. I., p. 97, Boston, 1857, in the following terms: "Catalogue raisonné, etc. I quote this paper under my name alone, because that of Mr. Desor, which is added to it, has no right there. It was added by him, after I had left Europe, not only without authority, but even without my learning it, for a whole year. . . . This is one of the most extraordinary cases of plagiarism I know of." Being the most important witness in the case, and the only survivor of all those who had anything to do with that "Catalogue," I shall dispose in a few words of the claims made rather cavalierly by Desor in his "Synopsis des Echinodermes fossiles," p. xv, Réponse à M. Agassiz, Paris, 1858.

Not only was I present many times when Agassiz dictated to Desor the descriptions of genera and species, and accompanied him often in his visits to the private and public collections of echinids in Paris, but it was to me that the manuscript was entrusted by Desor when he started for America, on the last day of February, 1847. About two-thirds of the "Catalogue" — the first eight sheets — had been printed under the supervision of Desor. I had to correct the proofs of sheets 9, 10, and 11, and besides to write not only the "Addenda," but also the entire "Distribution géologique des échinides fossiles," with many notes and corrections. The memoir was not issued in separate form until January, 1848, and it was I who delivered it to Agassiz at Cambridge, in May, 1848. I remember perfectly the amazement with which Agassiz saw the name of Desor on the cover as one of the authors, and as Agassiz knew the part I also had taken in the memoir, he said: "But you have more right than Desor to put also your name as one of the authors, for you did it entirely without compensation of any sort, only in kindness and friendship." On the whole, it was very presumptuous in Desor, who had assumed the position of *Maire du Palais*, ruling at his will, not only Agassiz's household, but also distributing scientific authorship according to his fancy or private interest. The part taken by him was simply that of a subordinate. Entirely in the pay of Agassiz, he simply wrote, mainly under Agassiz's dictation, the characteristics; added the description of about one hundred species — more or less — and three or four new genera, and also corrected a few errors, which was all a part of his duty as secretary. Agassiz had begun his studies and publications on the echinoderms five or six years before Desor came to Neuchâtel, and when he became Agassiz's secretary he knew absolutely nothing of echinoderms, or even of zoölogy.

The "Catalogue raisonné," etc., notwithstanding its many imperfections, marked great progress when it was published; and has, ever since, served as the basis of classification of the echinoderms. It is constantly quoted, and will continue to be quoted, just as the "Animaux sans vertèbres" of Lamarck is; and it is one of the great services rendered by Agassiz to zoölogy.

Agassiz was the recipient of all sorts of attention during his stay in Paris. He met many old friends, not only Parisians, but even men from the provinces and from foreign countries, who came to bid him farewell. M. Esprit Requien, the celebrated director of the museum at Avignon, who had communicated all his magnificent collection of fossil fishes, more especially those from the celebrated locality of Aix-en-Provence, for Agassiz's great monograph on the "Poissons fossiles," took lodging at the same hotel, the "Jardin du Roi," in order to see as much of Agassiz as possible. Requien was a rare type of savant: being an archæologist, a numismatologist, and a botanist and zoölogist, and a friend to every one with whom he came in contact, from Stendhal (Beyle), Prosper Mérimée, Adolphe Thiers, De Candolle, and Alcide d'Orbigny, to Agassiz. He possessed that exuberance of word and gesture so characteristic of the Provençal people and so well portrayed by one of their own writers, Alphonse Daudet. Agassiz much enjoyed his visit. There was another Provençal, Adolphe Thiers, who also was much attracted by the charm of Agassiz's society. They had previously met; but it was during Agassiz's present stay in Paris that a true friendship ripened between the two men, and their later correspondence showed many points of resemblance and common interest; both having an unbounded confidence in their power of conversation and public speech, and being extremely fond of applause and congenial society; they soon came to appreciate one another, and from this time Thiers, influenced by his conversations with Agassiz, became devoted to natural history. At that time, however, he had no leisure to give to it, being absorbed by his history of the Consulate and the Empire, and afterward by his political positions; but as soon as he was free after his Presidency of the third French Republic, he turned to science as a favourite study and consecrated the greater part of the last years of his life to the history of the earth.

There also came to Paris at this time, whether or not attracted by Agassiz it is impossible to say, one who had been a not over-scrupulous opponent of Agassiz on the glaciers, — no other than James D. Forbes, of Edinburgh, — and an attempt was made in his name to effect a reconciliation. After the publication by Agassiz, in 1842, of the history of his difficulties with Forbes, the scientific world, at least on the European continent, had pronounced against the method used by Forbes during and after his visits to the glacier of the Aar as Agassiz's guest. A common friend, Élie de Beaumont, invited Agassiz to a great dinner party to meet Forbes, insisting upon the desire on the part of Forbes to forget the past and be friends again; but Agassiz very politely, though firmly, declined the invitation, feeling that the attacks of Forbes had been marked by too great impropriety to allow of further friendly relations.

During his stay in Paris, it occurred to several of Agassiz's friends and acquaintances, that he might be induced to settle there permanently. Nothing would have been easier for the French government than to secure his services, if not at once, at least after his engagement with the Lowell Institute in Boston had been filled, and his promises to send collections to Berlin and Neuchâtel, in return for the advance money he had received from the king of Prussia, had been accomplished. For several reasons, the idea of his permanent residence in Paris was not to the taste of the leaders of natural history; although they feasted him, and gave him a Physiological prize of three hundred dollars at the annual meeting of the Institute of France, they feared, that if he became their colleague, he would soon over-shadow them all. In fact, jealousy was at the root of the affair; and although they loudly professed their admiration for the man himself and his work, and were ready to help him in some of his scientific work, they took no proper steps in the direction of keeping him. Nothing was offered in a direct way by the French government; but indirectly it was hinted that if he wanted to settle in Paris, official positions with salary amounting to six thousand francs per annum would be granted to him. Agassiz declined this doubtful offer, and it was probably a great relief to the official zoölogists and geologists to know that he was not to become their rival, and possibly their leader and master as well.

A Swiss artist of Neuchâtel, Fritz Berthoud, then a resident in Paris, took advantage of Agassiz's stay to obtain a full-length portrait of him. The picture, now in the museum at Neuchâtel, represents Agassiz and his secretary Desor; but the portrait of Agassiz is not good, and the picture, as a work of art, is poor, showing only the good will of the artist.

At the end of August, Agassiz left Paris, going first to London and then to Southampton, where he attended the meeting of the British Association for the Advancement of Sciences, the 10th of September. It was important for him to see Charles Lyell, who had lately returned from his two visits to North America, 1841–1842 and 1845, on June 26, and who had prepared the way for Agassiz, both with Mr. John A. Lowell, the director of the Lowell Institute at Boston, and with American savants in general, as to what might be expected from the visit of such a master and enthusiast in natural history.

During his short stay in England, Agassiz saw plainly that, although all the English leaders of sciences were extremely courteous and friendly to him, it was absolutely useless to expect from them the offer of any scientific position. His habit of going ahead, without regard to the consequences, was too much for English precision. They admired Agassiz; but that was all. Some, even, were ready to help him in a limited pecuniary way, and truly loved the savant, but the "sans-façon" of Agassiz they could not sanction.

At the end of September Agassiz embarked at Liverpool, on a steamer bound to Boston. The passage, as it is usually at about the time of the autumn equinox, was extremely rough; so much so that it was very much prolonged, and created apprehension as to the safety of the steamer. The newspapers even announced that the steamer was lost; and lamentations on the death of Agassiz were printed and circulated all through Switzerland: several of Agassiz's friends and admirers shed tears on reading the announcement of his tragical and premature death. "What a miserable end," says one of his best Swiss friends, "for poor Agassiz! He was much too valuable a savant to perish in the middle of the ocean."[1] Happily, the report was without foundation; but during the difficult crossing of the Atlantic Agassiz had full time to realize his position. He had left Europe much discouraged and in an extremely serious mood. During the past twenty years, he had acquired a great reputation, but he had had to pay very dear for it. Not only he had worked hard, and had even gone so far as to endanger his social position, but all his numerous publications had involved pecuniary losses, with the exception of the fishes of Martius and Spix of Brazil, and his two works now in the press in Paris, on the glaciers and the echinoderms. He had contracted debts which must be paid; and his position at Neuchâtel was on this account no longer tenable. Besides, he had formed the habit of having six, eight, and ten persons under his control, to help him in his works as assistants, secretary, artists, and lithographers. He had a family of three children to provide

[1] These two sentences may seem, now, rather melodramatic, but they well reflect the impression really produced. It must be remembered that, in 1846, the crossing of the Atlantic in steamships was in its infancy, many extremely serious accidents were then quite common, and steamers disappeared without leaving traces of any sort after them. Besides, in the centre of the continent, as Switzerland is, a journey to America was considered a great and dangerous undertaking.

for and an invalid wife whose health was a cause of great apprehension to all her friends. In addition, his stay in Paris and in England had dissipated all hope, if he had entertained any, of getting there official positions lucrative enough to satisfy his numerous wants and pecuniary obligations.

Success in America was for him a necessity, as he plainly saw, and he resolved to conquer, and bravely and nobly to meet his destiny, whatever came. The first thing for him to do was to master the English language sufficiently to allow him to speak in public and be understood. Ever since his first visit to England in 1834 he had practised more or less in translating and speaking English; but he knew very well, from his various attempts, how difficult it was for him to make himself understood among his English friends. Lyell had told him that it was useless to lecture in America in the French or German languages; for those two languages then were used in very narrow limits, and if he wished to make an impression on the American public, he must speak good English.

During his long journey across the Atlantic, Agassiz began in earnest, not only speaking English all the time, but committing to memory English sentences and repeating them aloud before any one who had the patience to hear him. The captain of the steamer said, "I have never had such a passenger as you, Professor Agassiz"; and like every one else, he was charmed with the great Swiss naturalist. Here again Agassiz's great memory helped him, although no longer so elastic as it had been in his youth; he soon knew a sufficient number of sentences and words to allow him to attempt public speaking, as we shall presently see. However, it was too late in life for him to become a complete master of the English language, as he was of the German. He never spoke correct English, and he always retained a strong French accent, which was not without some charm to his listeners.

CHAPTER XII.

1846 (*continued*)-1847.

ARRIVAL IN AMERICA, AND RECEPTION BY MR. JOHN A. LOWELL — CONDITION OF NATURAL HISTORY IN THE UNITED STATES — HIS FIRST VISIT TO NEW YORK — HIS ACQUAINTANCE WITH DR. SAMUEL MORTON, OF PHILADELPHIA — COLLECTIONS OF CAPTAIN WILKES MADE DURING HIS EXPEDITION ROUND THE WORLD, SEEN AT WASHINGTON — SCIENCE AT THE CAPITAL OF THE UNITED STATES — AGASSIZ'S FIRST SERIES OF LECTURES BEFORE THE LOWELL INSTITUTE AT BOSTON — HIS SUCCESS — A COURSE ON THE GLACIERS, IN FRENCH — FRANK DE POURTALÈS JOINS HIM — CHARLESTON, SOUTH CAROLINA — HIS OBSERVATIONS ON THE NEGROES — HIS DISAPPROVAL OF SLAVERY — ARRIVAL AT NEW YORK OF HIS TWO ASSISTANTS, DESOR AND GIRARD. — ESTABLISHMENT AT EAST BOSTON — SICKNESS OF AGASSIZ — HIS HOSPITALITY — A VISIT TO NIAGARA FALLS — ON BOARD THE UNITED STATES COAST SURVEY STEAMER, THE "BIBB" — ARRIVAL OF MINISTER CHARLES LOUIS PHILIPPE CHRISTINAT — FIRST DIFFICULTIES WITH HIS SECRETARY — TWO LETTERS TO J. MARCOU, EXTENDING AN INVITATION TO JOIN HIM.

ONE fine morning in the first week of October, 1846, a stranger recently disembarked was seen in the streets of Boston, looking to the right and left, in some astonishment, but steadily making his way to Pemberton Square, a rectangle with a garden in the centre, and surrounded by fine three-storied brick houses, at that time a very aristocratic part of the city, recalling many squares and circles of the London West End. After looking at the numbers of several houses, the foreigner pulled the bell at the door of Mr. John A. Lowell, who, on opening the door, was surprised to have a stranger, with a strong foreign accent, ask if Mr. Lowell was at home. The astonishment was quickly changed into undisguised satisfaction when the stranger added: "I—a-m P-r-o-f-es-s-or A-g-a-a-ss-i-z," with the drawling pronunciation so characteristic of Romand or French Switzerland, and more specially of Neuchâtel. Mr. Lowell very cordially extended both hands, and congratulated him on his safe arrival; and, in this auspicious manner, Agassiz made his entry into American life, and was launched into American society.

Lowell, with his keen eyes, his knowledge of European life and society, his association with savants, was very favourably impressed by Agassiz. He saw at once that his friends, Charles and Lady Lyell, had not overstrained the praise they had bestowed on the scientific worth of the savant they had so highly recommended to him; and from that first day he became an ardent supporter, and soon after a most intimate friend and counsellor, of Agassiz.

This day was certainly one of the happiest of Agassiz's life. A new life was opened to him at a moment of great mental depression and despondency, the natural result of the difficult position in which he was placed, both pecuniarily and socially.

The moment of his arrival in the New World was particularly fortunate and well timed. Until then the United States had developed without borrowing much from Europe. After the founding of the New England and Virginia colonies and the war of independence, American society, isolated and separated by the broad and stormy Atlantic, had been left to its own resources. At first a new society is necessarily limited to material progress, with sound moral and religious training; but sciences and the fine arts are not yet needed. Some scattered naturalists had here and there sprung up, but were not appreciated in proportion to their real merits, and were obliged to publish their observations in Europe, as was the case with the great ornithologist, Audubon. However, now that the great Napoleonic wars were over, a sort of revival in scientific researches and studies had begun. The American savants were not numerous enough to influence society; but a general desire to make scientific discoveries and to try what Americans could do for themselves in this field of human knowledge, illustrated by Buffon, Linné, Cuvier, Lamarck, de Candolle, etc., had already begun to exhibit signs of activity. Local scientific societies had sprung up at Philadelphia, Boston, New York, and Washington, and essays in scientific periodical publication, although not prosperous, because as yet a little premature, had shown that American savants, and especially American geologists, were desirous to enter the arena.

Curiously enough, science entered America led by geology. To be sure, botany, ornithology, conchology, entomology, and other branches of zoölogy, had some representatives scattered all along the Atlantic borders, and even as far west as New Harmony (then in the Indian Territory) in the Ohio Valley, but they were not only isolated, but also without the support of the people.

Public opinion did not encourage them. This was not the case with geology. People in general, and agriculturists in particular, soon showed an eager desire to know the resources of the soils, the rocks, and the mines. Geological surveys were started at the expense of the State in North Carolina, Virginia, Maryland, Pennsylvania, New Jersey, the New England States, New York, and Ohio. A desire to agree on points of classification and to know one another brought together the state geologists, who founded in 1840 the "Association of American Geologists," the first national scientific organization, and which held its meetings at different places in the Union.

The two visits of Lyell in 1841 and 1845, and the important journey of de Verneuil in 1846, among the palæozoic formations from the State of New York and Canada, to the Ohio Valley, the Upper Mississippi River, and Lake Superior, had given a strong impulse to geological researches, in bringing about the much needed comparison with European classification and synchronism. The field was well prepared, if not zoölogically, at least palæontologically, to receive one of the greatest palæontologists hitherto produced by Europe. The coming of Agassiz was anticipated with great joy by all American naturalists, and the more so, because at first his stay was announced to be only temporary.

After a few weeks spent in Boston, making the acquaintance of the Boston naturalists, and visiting the surrounding country, more especially the seashores and beaches, Agassiz went to New Haven, New York, Princeton, Philadelphia, Washington, and Albany. In this, his first experience, everything was new to him, — the people, the natural history, and American customs and society, — and his first impressions were most encouraging. With his extraordinary penetration and far-seeing vision, he realized what stores of scientific problems were in readiness, wanting only a little push to start the whole machinery of thorough researches over half a continent. It was just the work for him; American natural history had found its leader.

When I said that Agassiz was much encouraged by what he saw of American society, during October and November, 1846, it must not be understood that it was the fashionable world which he saw — rather limited although it was then, in comparison to what it is now. During the first five or six years of his life in America Agassiz paid very little attention to what is called fashionable society; he even avoided it, reserving his letters of introduction, and taking care to deliver them only at the last moment of his stay in New York and Washington, in order to escape invitations. His time was too precious to allow dissipation of any sort; so much so, that, on his first day in New York, instead of examining the magnificent bay and great city, he begged his cousin, Auguste Mayor, a resident of Brooklyn, to take him far up Greenwich Street, to the home of the only American palæo-ichthyologist, Mr. W. C. Redfield, and there he passed a part of the day, looking at fossil fishes.

His means did not allow him to go to first-class hotels, and he patronized second- and even third-class houses, or, more accurately, inns, as they were then, at Albany,

Philadelphia, and Washington. He readily adapted himself to American fare, except in one particular. Born in a wine country, even the excellent beer of Bavaria, during his long and numerous stays in Germany, was never much relished by him; and to be reduced to ice water and tea was rather hard. However, he was obliged often, too often for his inclination, to do the best he could, contenting himself with an occasional glass of claret, and a cup of black coffee, if obtainable, which was seldom the case. But as soon as he possessed a home, he provided light red wine and black coffee at luncheon and dinner, and adhered to this custom until the last day of his life. He never drank freely of strong wine, like the Spanish, Madeira, and Portuguese wines, and was averse to liquors of any sorts, excepting a small glass of "Chartreuse" or very old Cognac, when in company. Agassiz came to America too late in life to change this part of his diet.

At Princeton, Agassiz met, for the first time, Professor Joseph Henry, an American savant, who became one of his best friends and a constant admirer. Professor Asa Gray of Cambridge was there also, at the house of the then most celebrated botanist in the United States, Professor Torrey; and together Agassiz and Gray started for Philadelphia and Washington. Agassiz knew more of botany than was usual for a zoölogist; and Gray, then a young and rising botanist, was very solicitous to please Agassiz. Their friendship grew rapidly, until completely checked by the publication of Darwin's "Origin of Species," in 1859.

Philadelphia greatly attracted Agassiz. There he met

Dr. Samuel Morton, the great anthropologist, and an excellent palæontologist; a remarkable man, entirely to the taste of Agassiz, through the variety of his knowledge and the originality of his discoveries and thought. He also saw Conrad, Lea, Hallowell, Booth, and Frazer, and was, on the whole, well impressed by Philadelphian savants.

At Washington he was surprised by the gigantic scale on which the French engineer, Major Lenfant, had laid out the capital of the United States, by the imposing and beautiful Capitol, and also by the emptiness of many streets and quarters where building had hardly begun. It was, as it was called then, the "City of Magnificent Distances." Washington was not then the great and beautiful city of the present day. The inhabitants were few; and the government buildings, except the splendid Capitol, were limited to the White House, the State Department, the War and Navy buildings, and the Patent Office. The Smithsonian Institution existed only on paper;[1] and the savants were few in number, while the most prominent one, Professor Bache, the already celebrated director of the Coast Survey, was absent on duty. Arriving fresh from the great capital of France, it was a contrast to find science occupying so small a place in the great American republic, at least officially. Mr. Francis Markoe, the chief clerk of the State Department, and secretary of the National Institute, gave him a set of the Transactions of that society; and to the astonish-

[1] Professor Joseph Henry was not appointed secretary of the Smithsonian Institution until several months later, on the 3d of December, 1846.

ment of Agassiz, the three or four small paper-covered parts were far less important in regard to the quality, and even the number of original papers, than his "Bulletin de la Société d'Histoire Naturelle de Neuchâtel," issued in a very small town of one of the smallest cantons of Switzerland. The disappointment to one who, a few months before, under the dome of the Mazarin Palace, had received a Monthyon prize of physiology from the Royal Institute of France, may be easily understood. As a compensation, Markoe took Agassiz to the rooms of the Institute, and showed him the large and important collections made by Captain Wilkes during his scientific expedition round the world, from 1838 to 1842. He was more especially impressed by the extraordinarily beautiful and exact drawings of fishes, reptiles, molluscs, and corals, executed from life during the expedition by Mr. Drayton, by far the best artist of natural history objects in America.

Until this time, all exploring expeditions into the interior of the United States, sent at the expense of government, from the journeys of Lewis and Clarke, Pike, Major Long, Nicolet, Featherstonhaugh, D. D. Owen, to those of Captain Fremont, had had their reports rather meagrely published in regard to plates and natural history drawings. Congress always voted liberal sums to defray the expense of these publications, but they were at that time all done by contract, falling as spoils into the hands of politicians; and the result was the issue of reports disgraceful as regards material execution — bad type, bad drawings, bad paper — a state of things most discouraging to all the explorers and savants connected with the government.

In order to remedy such a condition, all the reports of the Wilkes Exploring Expedition were placed under the direction of one man, Mr. Drayton, who superintended the whole publication. But, going from one extreme to another, the Senate, which made the law, inserted in it a provision by which the number of copies of each volume was limited to two hundred, and distributed exclusively to senators; while of Captain Fremont's report, issued in 1845, ten thousand extra copies were printed for the use of the members of Congress. The immense difference between two hundred and ten thousand copies is evident. The result was that Wilkes's reports, being placed exclusively in the hands of senators, no one of whom was a scientific man, or had sufficient knowledge of natural history to appreciate their value but distributed them simply on account of the beautiful plates, became extremely rare from the very first. Half of the number of copies was soon entirely lost, and some of the reports were destroyed in a fire at the printing establishment, so that now several of the great quarto volumes and folio atlases of the expedition have become so scarce that it is almost impossible to get copies at any price.

The report of Fremont, which was defective only in good execution, was furnished with poor engravings, poor plates of fossils, poor paper, and printed from indifferent type. When Agassiz received at Washington, from the hands of Colonel Abert, chief of Topographic Engineers, Fremont's report and those of Nicolet, Abert's son, D. D. Owen, and Featherstonhaugh, he immediately saw that a great reform was needed to give their true value to all these government reports and publications. On the one hand, Wilkes's report was lost to the scientific public by its scarcity and the mode of distribution; and, on the other hand, Fremont's report and others of the same sort were so badly executed that they were a disgrace to the country.

From this moment Agassiz began to urge constantly on those in power at Washington the necessity laid upon the United States government to publish only well-executed volumes, especially in regard to plates of natural history and landscape drawings. He himself set the example in 1850 in publishing his important exploration of Lake Superior. His efforts, combined with the powerful help of Professor Bache and of Professor Henry, succeeded in bringing about a much better state of things after 1853, as we shall see. But it was during his first visit to Washington, in 1846, that he laid the foundation for the improvement of the government scientific publications.

As soon as Agassiz was back in Boston, he again devoted himself to his practice of learning English phrases by heart, and speaking aloud in English in his room in order to be able to deliver his first course of lectures before the Lowell Institute. The subject was "The Plan of the Creation, especially in the Animal Kingdom." It is not surprising that he was much concerned about his first lecture at the beginning of December; for it was not an easy task to set forth, in a language which he had never before used in public, one of the most difficult and complicated questions of natural history, but he was so full of his subject that he trusted to his power to enrapture his very large audience of fifteen hundred persons of both sexes and of all ages. Sometimes words were not at his command, and he would pause and wait patiently, with his peculiar smile and beaming eyes, so characteristic of the man, in the meantime amusing his audience by drawing on the blackboard excellent outlines of animals. His French accent was considered a new charm added to his other personal accomplishments; and he stepped down from the platform in a burst of applause, which plainly showed that he had succeeded in his rather hazardous undertaking.

Until then he had never seen a scientific lecture delivered before so many people. The largest audiences he had seen were in Paris at the lecture-rooms of the Collége de France and the Jardin des Plantes, when George Cuvier was the lecturer, and at the Astronomical Observatory, when François Arago was explaining the "Systême du Monde" before such listeners as Alexander von Humboldt, Biot, Leverier, and a whole crowd of members of the French Institute. In those days three or four hundred persons at most crowded the Paris lecture-rooms, but the fifteen hundred auditors of the Lowell Institute room surpassed everything he had ever thought of. Making a large allowance for the curiosity which attracted many persons, there remained enough to satisfy, and even more than satisfy, his most sanguine expectations. For the first time he understood that very characteristic feature of American life, — public lectures. He was impressed by the seriousness of his listeners, although he knew well that only a small part of the audience was able to understand the full meaning of what he said; but it was very encouraging to see so many ladies and gentlemen of the world ignorant, almost all of them, of the first elements of natural history, listening attentively to what he had to say. It showed a desire to learn, or at least to be instructed on points in regard to which very few of them before entering the lecture-room had the least knowledge. It was a revelation to him, which from that day caused a great change not only in his scientific life, but also in his social and family habits.

It is fortunate for the progress of science, to which Agassiz contributed so largely during his twenty years of work in Europe, that he did not begin his scientific life in America, for his extraordinary ability as a teacher would have absorbed all his time. To be sure, he would have popularized natural history, by a constant contact, of forty-five years' duration, with the general mass of the American people; but he would never have undertaken his "Poissons fossiles," and many other of his original works. Although his first course of lectures in America, at the Lowell Institute, was a success, Agassiz felt that a part of his power was paralyzed, in a great degree, by the difficulty he experienced in using the English language. For a man who was a good scholar in Latin, in Greek, in French, and in German, it was painful to realize how incorrect his English was, and it was a great regret to him not to be able to display all his resources and his unequalled talent as a teacher "hors ligne."

His friends in Boston and Cambridge understood this feeling, and, at their request, Agassiz delivered, before a select audience, a series of lectures on "Les glaciers et l'époque glaciaire," in French, his native language. At that time, the number of persons in Boston and Cambridge who knew enough French to follow a lecture in that language was limited. However, the subscription list was large, the ladies outnumbering the gentlemen, and according to his own account it was the best course of lectures he ever delivered. The subject was entirely new in America; the illustrations were excellent and most attractive for the time, and the delivery in correct and even elegant French. It was a rare treat to every one, from the lecturer himself to almost all his listeners, the most enthusiastic being the ladies, who were lost in admiration of the Alpine glaciers, Alpine peaks, Jura boulders, "roches moutonnées," and "cailloux striés," and, indeed, of the Professor.

After these two courses of lectures, Agassiz became a great favourite in Boston society, and he remained such until the end of his life. He had conquered the "élite" of Boston and Cambridge, as well as the common people, not only of Boston, but of Massachusetts and even of New England; for his lectures were published at once, and almost *in extenso* in newspapers.

During the delivery of his Boston lectures, his favourite pupil at Neuchâtel, Frank de Pourtalès, had joined him, the first of Agassiz's European scientific friends to come to this country, attracted by his glowing accounts in his private letters from America. The addition of Pourtalès, who had independent means, was important, for Agassiz did not have to provide for his support, and he was greatly assisted by him, when he settled at East Boston.

After repeating his Lowell lectures at Albany, before a very sympathetic audience, Agassiz and Pourtalès embarked at New York for Charleston, South Carolina. The reception they received was particularly gratifying. Everything possessed a charm unknown to Agassiz until then, and it was the first time that he came in contact with a sub-tropical fauna and flora. Besides, the broad and generous hospitality of the planters attracted him much, and Agassiz and Pourtalès were both glad to meet gentlemen, coming from their common stock of French and Swiss Protestants, like de Saussure, Ravenel, and others, or Dr. Fabre, an old Swabe student of the University of Tübingen. But the man who particularly pleased them was Dr. Holbrook, a herpetologist of talent, one of the rare zoölogists of the New World, and at the same time a most amiable and serviceable man.

Agassiz delivered a course of lectures, with the same success as before the Lowell Institute, which made him at once a great favourite in Southern society. Established with Pourtalès on one of the islands near Charleston, he was in perfect ecstasy over his daily discoveries of new fishes, new turtles, new molluscs. The rich entomological fauna was also a constant surprise. But what made the greatest impression on him as a naturalist was his contact with a large population of negroes. With his power of comparing zoölogical characters, it was impossible for him to consider the black man as a species identical with the white man. To one who considered not only the species, but even the genus, as natural divisions, whatever the system of classification adopted, the conclusion was irresistible.

One of his last lectures, just before leaving Neuchâtel, was on the geographical distribution of animals ("Notice sur la géographie des animaux," "Revue Suisse," avril, 1845), in which he had insisted that every animal and plant is confined to a certain portion of the earth, while man is the only one which covers the whole surface. As he says, "L'homme, malgré la diversité de ses races, constitue une seule et même espèce sur toute la surface du globe." It was hard for him to abandon this view; but he was too thorough a naturalist, and had a too exalted idea of the immutability of species, like his master, Cuvier, to believe in only races for man. After his first visit to South Carolina, species, in his eyes, existed for man as well as for every other genus. That is to say that the genus *homo* is composed of several species; for instance, the Caucasian or white man is one species, with many varieties or races, such as the Arabs, Indians, Turks, Scandinavians, Irish, Slavic, Greeks, Italians, etc. The negro is another species with many races or varieties, such as the Hottentots, the Soudans, the Congos, the Zambesi, etc. But it would be erroneous to conclude, from his opinion as a naturalist, that he was in favour of slavery. This was an abyss which he never crossed. The passionate and bitter discussions, which already agitated and divided the South from the North, had no influence on him, and he never took part in them, directly or indirectly. It is true that several politicians of the time made use of his opinions for their own selfish interests, but it was impossible for Agassiz to prevent it. Confining himself to a zoölogical point of view, he admitted with great sincerity and frankness, that although once a believer in the unity of the races of man, he had found out that this was an error, and that his studies among large numbers of negroes and Indians had led him, as a zoölogist, to conclude that it was impossible to consider them as simple varieties or races of the white man. In his view, they were entirely distinct species, each, — negroes, American Indians, and Circassians or Europeans, — possessing its peculiar varieties or races.

But as regards the servitude of one species to another, and the right of one man to sell another, Agassiz never, for an instant, justified such a proceeding, either morally, socially, or religiously. Science had nothing to do with such an iniquity; to deal with it was the work of morality, philanthropy, politics, and religion, but not of a savant, whose domain is entirely outside of all institutions of society.

In early spring Agassiz returned to New York, where he met his assistants, Edward Desor and Charles Girard, who had left Paris in February, and had embarked on a sailing-ship at Havre, the 2d of March, 1847. It now became needful to have a permanent establishment somewhere; and Agassiz did not hesitate to choose Boston as his headquarters, on account of the great interest and sympathy shown to him since the day of his arrival on American soil; and, curiously enough, the house he leased was only a stone's throw from his landing-place at the Cunard wharf.

Accompanied by Pourtalès, Desor, and Girard, he came to Boston, early in April, stopping at a boarding-house in Temple Place, preparatory to arranging for a house. Agassiz took, for one year, a three-storied brick house at East Boston, close by the sea, the tide even entering the garden; where he tied up a little row-boat, called, in New England, a dory, as his first contribution to the furniture of his establishment. Here is another example of atavism, in a descendant of the lake-dwelling peoples of Switzerland, who were always ready to return to water, whenever occasion offered. He was led to the choice of this house, with its rather heavy rent, — one thousand dollars a year, — by his ardent desire to have a laboratory close by the sea, where he could get marine animals to his heart's content, and preserve them alive.

It was not easy for four Europeans, three of whom spoke hardly a word of English, to furnish a house, and remove there all their property, including books, large diagrams, and the several barrels and boxes of natural history specimens collected since Agassiz's arrival.

Before the final arrangement and the removal to East Boston, the health of Agassiz broke down, for the first time in his life. Until then fatigue and anxiety of all sorts had made no impression on his strong constitution; he seemed to be above the reach of sickness. But the numerous exertions on many lines entirely different from those to which he was accustomed; the American way of living, so new to him, added to his great anxiety as to his future, which was still uncertain; all this fell heavily upon him; and it is not surprising that a few days after his return to Boston he was seized by a severe attack of nervous prostration, a malady which clung to him from this time to the end of his life, recurring now and then, with an increase in the frequency of its attacks as he grew older, and as he constantly and often imprudently burdened himself with new duties.

By the end of May the settlement was achieved; rooms were assigned for microscopical studies, for the dissection of animals, for the drawing of large diagrams for public lectures, and the collections were sorted and divided for future distribution. Every day Pourtalès and Charles Girard went sailing in Boston harbour, dredging the bottom for specimens; or they followed on foot the edge of the tide water on beautiful Chelsea beach, picking up every animal worth preserving.

The originality of this naturalist-home brought to East Boston not only all those engaged in the study of natural history, but also many ladies and gentlemen curious to see how practical zoölogy could be made. Agassiz, with his usual buoyancy of spirit, and his ever-ready desire to teach, showed the ladies how to look into the microscope, explaining graphically the wonders of each small animal. Then, turning to the tank of salt water always teeming with marine animals, he would take a fish, or a big jellyfish and explain its way of swimming, or its system of blood circulation. Time passed quickly, and his visitors left him charmed with what they had heard and seen. Boston felt proud of the acquisition of a naturalist of genius, while Agassiz was delighted to have excited an interest among persons so intelligent and refined in taste.

During the heat of summer, Mr. Lowell, always attentive to the comfort and welfare of Agassiz, invited him and his assistant, Desor, as his guests, to visit Niagara Falls and the great rapids of the St. Lawrence River. The impression of this grand and picturesque region, combined with the finding of glacial scratches everywhere, and the sight of many zoölogical specimens, especially fishes, created in Agassiz an admiration and an enthusiasm difficult for any one not a naturalist to realize, and from that moment he was resolute to consecrate the remainder of his life to the study of the natural history of the New World.

Returning to Boston, he received an invitation from Professor Bache to join in a cruise along the shores of Cape Cod and the island of Nantucket, on the coast-survey steamer *Bibb*, commanded by Lieutenant (afterward Admiral) Charles Henry Davis, U. S. Navy, who was then employed in surveying the bay of Boston,—an excursion which passed for Agassiz like a dream of the Thousand and One Nights. In one day, as he says, he learned more than in months from books or dried specimens. It was a new opening for his never-ending activity of spirits and schemes. A most intimate friendship grew up with both Professor Bache and Lieutenant Davis from that first cruise, and lasted as long as they lived, and in them Agassiz found, not only sympathizers, but true patrons of scientific researches, happy in the opportunity to secure to America the services of such a savant. There is no doubt that Agassiz's settlement in America was due to the kind reception and many acts of true friendship and admiration he received from Mr. Lowell and Professor Bache.

Before his first year in America was over, a most intimate friend of his Swiss family, M. Charles Louis Philippe Christinat, arrived at his house in East Boston. Victim of a political revolution in the Canton de Vaud, Christinat, for many years a minister in the village of Montpreveyres, was obliged to leave his parish, and after wandering as an exile in Italy and France, he resolved to join his friend Agassiz, and finish his life with him. He possessed the full confidence of Agassiz's mother, and the family was very glad that such a trusty friend was willing to help Agassiz by his advice and his devotion to his person and interests; for they all knew how much Agassiz was influenced, and often not in the right direction, by his secretary Desor.

As soon as Christinat arrived, at the end of September, 1847, Agassiz, who remembered how devoted Christinat had always been to him since his childhood, going so far as to supply his always empty pocket with money in order that he might make his much-desired journey to Paris, felt that he had at last near him a man whom he could fully trust. It was a great relief to his mind. His relations with Desor were no longer as friendly as they had formerly been at Neuchâtel. When they met again in April, after an eight months' separation, Agassiz saw at once a great change in Desor's manner, and more especially in his way of talking. He had left him in Paris his secretary and assistant, and he found him at New York his associate and collaborator, with a certain air of domination which extended even to every act of his private life. Passionate and painful discussions followed one another in rapid succession; and although they all ended in reconciliation, they were but the beginning of most serious difficulties. It was evident that Desor's prolonged sojourn at Paris, during which he had assumed the joint authorship of one of Agassiz's publications, and his journey in Scandinavia—at the expense of Agassiz, who found the amount of one thousand dollars a little hard to pay back to his banker in his already straitened pecuniary position—had given him a somewhat exalted opinion of his scientific and social value. Agassiz was much hurt by this new demeanour of his secretary; it was hard for him to be lectured by his own pupil both on scientific and private affairs. He recalled the poor young man who came to him at Neuchâtel at the end of 1837, not as a naturalist of worth, but only as an amanuensis and translator, and at whose mercy now, ten years later, he found himself, both scientifically and socially. As he himself said, it was he who brought the water to turn the mill, for Desor had never contributed a cent to the constantly increasing expenses.

The following letters are presented to show how Agassiz was always ready to help and encourage a young naturalist; and they allow me, at the same time, to define my position with him. Being an assistant at the Jardin des Plantes, under the direction of M. Cordier, the professor of geology, I was offered, as a reward for work done in determining invertebrate fossils at the Museum, a journey of three years' duration, outside of Europe, with my own choice as regarded the country to be explored. My acquaintance with Agassiz led me to choose North America, and I wrote him asking if he would help me by his advice, and tell me his plans for explorations during 1848. His answer follows: —

BOSTON, 30 septembre, 1847.

MONSIEUR JULES MARCOU,
Paris.

Mon cher monsieur, — Ne pouvant écrire aujourd'hui à M. Cordier, ni vous donner de quelques semaines une esquisse arrêté de mes projets de voyage pour l'année prochaine et ne voulant cependant pas vous laisser attendre une réponse à la demande que vous m'adressez de venir rejoindre aux États-Unis, le trio de travailleurs que vous avez connu à Paris, je me bornerai pour le moment à vous dire en termes généraux que je serai charmé de vous associer à ce que je puis faire dans ce pays. Je sais trop bien tout ce qu'il reste à faire dans tous les domaines de la science pour redouter le concours d'efforts combinés dans un même but ; bien au contraire je crois que les résultats scientifiques, que notre petite troupe pourra obtenir seront d'autant plus considérables qu'elle s'associera un plus grand nombre de bons observateurs ; et comme je n'ai aucun penchant à m'approprier les observations d'autrui, vous pouvez être assuré d'avance que quelqu'importantes ou quelqu'insignifiantes que puissent être les découvertes que vous ferez dans ces pérégrinations communes, elles vous seront bien duement acquises, et vous resteront en plein et sans partage, même dans le cas où elles auraient été amenées par des recherches que j'aurais pu vous suggérer. C'est sur de telles bases seulement que je conçois des rapports durables entre hommes dévoués à la science.

Quant à mes projets prochains, j'ai l'intention de visiter cet hiver les Carolines et de revenir pour le mois de février à Boston, pour me préparer à une course dans l'Ouest. Je désire consacrer une bonne partie de l'été à l'exploration des bords du Lac Supérieur et de la vallée du Mississippi, comme préparation à une seconde course au-delà de ce fleuve, dans la direction des Montagnes Rocheuses. Les ressources dont je pourrai disposer ne me permettent pas de songer à passer l'hiver dans une sorte d'inaction, loin des grandes villes, où je puis par quelques leçons acquérir de quoi poursuivre mes recherches. Je crois de plus qu'il est plus avantageux de couper ainsi en deux temps une exploration de l'Ouest dont la première campagne servira de reconnaissance et de point de départ pour la seconde. Puis il y a dans l'état de l'Ohio plusieurs collections qui méritent d'être étudiées et dont l'examen nous évitera des travaux inutiles, et nous fournira des points de repair. Même ici à Boston, mais surtout à Albany, vous pourriez consacrer bien des mois utilement à vous préparer, car il vous sera difficile, malgré les publications des géologues de New York, de vous faire une juste idée de l'étendue des travaux, en grande partie inédits, qui ont été faits dans ces contrées. Dans ce moment deux caravanes de géologues explorent les états de Michigan et de Wisconsin. Aussi plutôt vous pourrez venir et mieux, et surtout ayez à votre disposition des ressources pécuniaires suffisantes, car pour la dépense les dollars sont à peu près pour nous ce que les francs étaient à Paris, avec un genre de vie qui est à peu près le même.

Ces renseignements préliminaires vous permettront de faire vos préparatifs sans délai; dès que je le pourrai, je vous écrirai d'une manière plus précise à quoi je compte m'arrêter définitivement pour l'année prochaine, et j'écrirai en outre à M. Cordier pour l'assurer que les intérêts du Muséum ne courront aucun risque si vous venez me joindre. Je n'en veux de meilleure preuve que le fait que j'ai déjà mis spontanément de côté une assez jolie collection de fossiles paléozoiques que je destine au Jardin des Plantes et qui seraient déjà partis pour l'Europe n'était l'ennui de l'emballage.

Je vous remercie de tous les soins que vous avez mis à l'impression des Echinodermes (Catalogue raisonné) et à la rédaction d'un registre de la distribution géologique, addenda, etc.; ce sont des additions qui seront très utiles, je pense.

Adieu, mon cher Monsieur; croyez à la sincérité de l'intérêt que je vous porte et que votre zèle pour la géologie justifie si complétement.

Votre dévoué,

LS. AGASSIZ.

In another letter, dated New York, 14 November, 1847, he says: —

Desor est resté à Boston pour le moment; Charles, dont je suis assez content, est ici avec moi; Pourtalès m'accompagne dans le Sud avec mon dessinateur. Nous ferons de la bonne besogne, je crois, et malgré la difficulté de gagner ma vie en faisant des cours depuis que je suis au terme de mes subsides de Berlin, tout va son train, comme devant, et je suis loin de songer à réduire l'étendue de mes recherches pour peu que je puisse continuer à suffire à toute cette dépense à force de travail. Adieu, mon cher ami, venez bien vite et je crois que vous serez content de ce pays.

Tout à vous,

LS. AGASSIZ.

END OF VOLUME I.

Printed in the United States
By Bookmasters